D0974959

KEPLER'S CONJECTURE

How Some of the Greatest Minds in History
Helped Solve One of the Oldest
Math Problems in the World

George G. Szpiro

WILEY

John Wiley & Sons, Inc.

Published by John Wiley & Sons, Inc., Hoboken, New Jersey
Published simultaneously in Canada

Illustrations on pp. 4, 5, 8, 9, 23, 25, 31, 32, 34, 45, 47, 50, 56, 60, 61, 62, 66, 68, 69, 73, 74, 75, 81, 85, 86, 109, 121, 122, 127, 130, 133, 135, 138, 143, 146, 147, 153, 160, 164, 165, 168, 171, 172, 173, 187, 188, 218, 220, 222, 225, 226, 228, 230, 235, 236, 238, 239, 244, 245, 246, 247, 249, 250, 251, 253, 258, 259, 261, 264, 266, 268, 269, 274, copyright © 2003 by Itay Almog. All rights reserved

Photos pp. 12, 37, 54, 77, 100, 115 © Nidersächsische Staats- und Universitätsbibliothek, Göttingen; p. 52 © Department of Mathematics, University of Oslo; p. 92 © AT&T Labs; p. 224 © Denis Weaire

For general information about our other products and services, please contact our Customer Care Department within the United States at (800) 762-2974, outside the United States at (317) 572-3993 or fax (317) 572-4002.

Wiley also publishes its books in a variety of electronic formats. Some content that appears in print may not be available in electronic books. For more information about Wiley products, visit our web site at www.wiley.com.

Library of Congress Cataloging-in-Publication Data:

Szpiro, George, date.
    Kepler's conjecture : how some of the greatest minds in history helped solve one of the oldest math problems in the world / by George Szpiro.
        p. cm.
    Includes bibliographical references and index.
    ISBN 0-471-08601-0 (cloth : acid-free paper)
    1. Mathematics—Popular works.   I. Title.
    QA93 .S97 2002
    510—dc21

                                                                                2002014422

Printed in the United States of America

10 9 8 7 6 5 4 3 2 1

Contents

Preface

This book describes a problem that has vexed mathematicians for nearly four hundred years. In 1611, the German astronomer Johannes Kepler conjectured that the way to pack spheres as densely as possible is to pile them up in the same manner that greengrocers stack oranges or tomatoes. Until recently, a rigorous proof of that conjecture was missing.

It was not for lack of trying. The best and the brightest attempted to solve the problem for four centuries. Only in 1998 did Tom Hales, a young mathematician from the University of Michigan, achieve success. And he had to resort to computers. The time and effort that scores of mathematicians expended on the problem is truly surprising. Mathematicians routinely deal with four and higher dimensional spaces. Sometimes this is difficult; it often taxes the imagination. But at least in three-dimensional space we know our way around. Or so it seems. Well, this isn't so, and the intellectual struggles that are related in this book attest to the immense difficulties. After Simon Singh published his bestseller on Fermat's problem, he wrote in *New Scientist* that "a worthy successor for Fermat's Last Theorem must match its charm and allure. Kepler's sphere-packing conjecture is just such a problem—it looks simple at first sight, but reveals its subtle horrors to those who try to solve it."

I first met Kepler's conjecture in 1968, as a first-year mathematics student at the Swiss Federal Institute of Technology (ETH). A professor of geometry mentioned in an unrelated context that "one believes that the densest packing of spheres is achieved when each sphere is touched by twelve others in a certain manner." He mentioned that Kepler had been the first person to state this conjecture and went on to say that together with Fermat's famous theorem this was one of the oldest unproven mathematical conjectures. I then forgot all about it for a few decades.

Thirty years and a few career changes later, I attended a conference in Haifa, Israel. It dealt with the subject of symmetry in academic and artistic disciplines. I was working as a correspondent for a Swiss daily, the *Neue Zürcher Zeitung* (*NZZ*). The seven-day conference turned out to be one of the best weeks of my journalistic career. Among the people I met in Haifa was Tom Hales, the young professor from the University of Michigan, who had just a few weeks previously completed his proof of Kepler's conjecture. His talk was one of the highlights of the conference. I subsequently wrote an article on the conference for the *NZZ*, featuring Tom's proof as its centerpiece. Then I returned to being a political journalist.

The following spring, while working up a sweat on my treadmill one afternoon, an idea suddenly hit. Maybe there are people, not necessarily mathematicians, who would be interested in reading about Kepler's conjecture. I got off the treadmill and started writing. I continued to write for two and a half years. During that time, the second Palestinian uprising broke out and the peace process was coming apart. It was a very sad and frustrating period. What kept my spirits up in these trying times was that during the night, after the newspaper's deadline, I was able to work on the book. But then, just as I was putting the finishing touches to the last chapters, an Islamic Jihad suicide bomber took the life of one my closest friends. A few days later, disaster hit New York, Washington, and Pennsylvania. If only human endeavor could be channeled into furthering knowledge instead of seeking to visit destruction on one's fellow men. Would it not be nice if newspapers could fill their pages solely with stories about arts, sports, and scientific achievements, and spice up the latter, at worst, with news on priority disputes and academic battles?

This book is meant for the general reader interested in science, scientists, and the history of science, while trying to avoid short-changing mathematicians. No knowledge of mathematics is needed except for what one usually learns in high school. On the other hand, I have tried to give as much mathematical detail as possible so that people who would like to know more about what mathematicians do will also find the book of interest. (Readers interested in knowing more about the people who helped solve Kepler's conjecture and the circumstances of their work will also be able to find additional material at www.GeorgeSzpiro.com.)

Those readers more interested in the basic story may want to skip the more esoteric mathematical points; for that reason, some of the denser mathematical passages are set in a different font. Even more esoteric material is banished to appendixes. I should point out that the mathematics is by no means rigorous. My aim was to give the general idea of what constitutes a mathematical proof, not to get lost in the details. Emphasis is placed on vividness and sometimes only an example is given rather than a stringent argument.

One further math note: throughout the text, numbers are truncated after three or four digits. In the mathematical literature this is usually written as, say, 0.883. . . . , to indicate that many more digits (possibly infinitely many) follow. In this book I do not always add the dots after the digits.

I have found much valuable material at the Mathematics Library, the Harman Science Library and the Edelstein Library for History and Philosophy of Science, all at the Hebrew University of Jerusalem. The library of the ETH in Zürich kindly supplied some papers that were not available anywhere else, and even the library of the Israeli Atomic Energy Institute provided a hard-to-find paper. I would like to thank all those institutions. The Internet proved, as always, to be a cornucopia of much useful information . . . and of much rubbish. For example, under the heading "On Johannes Kepler's Early Life" I found the following gem: "There are no records of Johannes having any parents." So much for that. Separating the e-wheat from the e-chaff will probably become the most important aspect of Internet search engines of the future. One of the most useful web sites I came across during the research for this book is the MacTutor History of Mathematics archive (www-groups.dcs.st-and.ac.uk/~history), maintained by the School of Mathematics and Statistics of the University of Saint Andrews in Scotland. It stores a collection of biographies of about 1,500 mathematicians.

Friends and colleagues read parts of the manuscript and made suggestions. I mention them in alphabetical order. Among the mathematicians and physicists who offered advice and explanations are Andras Bezdek, Benno Eckmann, Sam Ferguson, Tom Hales, Wu-Yi Hsiang, Robert Hunt, Greg Kuperberg, Wlodek Kuperberg, Jeff Lagarias, Christoph Lüthy, Robert MacPherson, Luigi Nassimbeni, Andrew Odlyzko, Karl Sigmund, Denis Weaire, and Günther Ziegler. I thank all of them for their efforts, most of all Tom and Sam, who were always ready with an e-mail clarification to any of my innumerable questions on the fine points of their proof. Thanks are also due to friends who took the time to read selected chapters: Elaine Bichler, Jonathan Dagmy, Ray and Jeanine Fields, Ies Friede, Jonathan Misheiker, Marshall Sarnat, Benny Shanon, and Barbara Zinn. Itay Almog did much more than just the artwork by correcting some errors and providing me with numerous suggestions for improvement. Special acknowledgment is reserved for my mother, who read the entire manuscript. (Needless to say, she found it fascinating.) I would also like to thank my agent, Ed Knappman, who encouraged me from the time when only a sample chapter and an outline existed, and Jeff Golick, the editor at John Wiley & Sons, who brought the manuscript into publishable form.

Finally, I want to express gratefulness and appreciation to my wife, For-tunée, and my children Sarit, Noam, and Noga. They always bore with me when I pointed out yet another instance of Kepler's sphere arrangement. Their good humor is what makes it all worthwhile. This book was written in no little part to instill in them some love and admiration for science and mathematics. I hope I succeeded. My wife's first name expresses it best and I want to end by saying, *c'est moi qui est fortuné de vous avoir autour de moi!*

This book is dedicated to my parents, Simcha Binem Szpiro (from War-saw, Poland) and Marta Szpiro-Szikla (from Beregszasz, Hungary).

CHAPTER 1

Cannonballs and Melons

The English nobleman and seafarer Sir Walter Raleigh (1552–1618) is perhaps an unlikely progenitor for an intellectual adventure. His scholarly achievements somewhat in doubt, he nevertheless set in motion one of the great mathematical investigations of the past four hundred years: Sometime toward the end of the 1590s, stocking his ships for yet another expedition, Raleigh asked his sidekick and mathematical assistant Thomas Harriot to develop a formula that would allow him to know how many cannonballs were in a given stack simply by looking at the shape of the pile. Harriot, no slouch, solved the problem put to him by Raleigh. But understanding his master's needs like any good assistant, he took it a step further and attempted to discover the most efficient way to stuff as many cannonballs into the hold of a ship as possible. And thus a mathematical problem was born.

Harriot, Sir Walter's junior by eight years, was an accomplished mathematician, astronomer, and geographer. He was also an ardent atheist, a persuasion that he shared with his master but that was not to be flaunted. The two men had been introduced to each other by a common tutor and their shared interest in navigation and exploration was the basis for a lifelong friendship.

One of Harriot's few surviving written documents is his report on Sir Walter's expedition of 1585–1586 to the New World: *A Briefe and True Report of the New Found Land of Virginia*. Published in 1588, it was the first English book describing the first English colony in America. The report became quite a hit with the literati of the time, was reprinted several times, and was translated into Latin, French, and German. Because of this report, Harriot is better remembered as an observer of the American way of life than as a scientist.

Harriot's scientific achievements are many, although he is sometimes quite unjustly overlooked as one of the foremost thinkers of his time. In 1609, Harriot was the first man to look at the moon through a telescope, and he discovered sunspots and the moons of Jupiter independently of Galileo. This we know only from his notebooks, however, because Harriot hardly published anything. Most of his scientific findings are contained in his magnum opus, *Artis analyticae praxis ad Aequationes Algebraicas Resolvendas* (Applications of the art of analysis to the solution of algebraic equations), published in 1631, ten years after his death. In this book, Harriot developed a numerical method to approach solutions of algebraic equations. He also advanced the techniques to solve equations of the third degree, and is credited with introducing the signs ">" (greater than) and "<" (less than) into mathematical notation. He contributed to the understanding of the refraction of light, binary mathematics, spherical geometry, ballistics, and many other fields. In 1607 he observed a UFO in the night sky that would later be identified as Halley's comet. He was also one of the first atomists (thinkers who were convinced that all matter is made up of minute particles), at a time when this view was not at all popular. And he had all the insights into crystalline order that were later attributed to the more famous astronomer Johannes Kepler.

In answering Sir Walter's question, Harriot devised a table that helped determine the number of cannonballs on carts of given shapes. But as mentioned, Harriot went one step further. Not only did he devise formulas to compute how many cannonballs were in stacks of a certain shape, but he would also discover how to maximize the number of cannonballs that would fit in the hold of a ship. In modern mathematical parlance, he wondered how three-dimensional spheres could be packed as densely as possible. After contemplating the question for a while, Harriot decided to write a letter to one of the foremost mathematicians, physicists, and astronomers of the time—Kepler, his colleague in Prague.

Although cannonballs are three-dimensional objects, the same problem can also be formulated in lower dimensions, and we will first have a look at the corresponding problem in one dimension and two dimensions. The objects that interest us are *spheres,* which we define formally as the collection of all points in space, whose distance to the center is smaller or equal to a certain radius. Space and distance are defined with respect to their dimension. In one dimension, space is just a line. In two dimensions, space is a surface. And three-dimensional space is the space all around us. So, according to the definition, a one-dimensional sphere is just a piece of a line with length twice its radius. To make this a bit more intuitive, look at a line and decide on a certain point as the center of the sphere. Then move first in one direction along the line until you have covered a distance of R, and then do the same in the other direction. This is the one-dimensional sphere of radius R. It may seem surprising at first that a straight line can be a sphere, since we usually think of spheres as round objects.[1] But that should not bother us; "roundness" has no meaning in one dimension.

A two-dimensional sphere is a more familiar object. Define a point in the plane and then move a distance of R in all directions; this sphere consists of the circle and of all the points inside it. You can picture it in the following way: Imagine a field with a pole in the middle. Attach a cow to the pole with a rope of length R, and let it graze. After a while the cow will have grazed off the grass at all the points that are no farther away from the center than radius R.

Finally, the three-dimensional sphere is, of course, our cannonball.

Why stop at three dimensions? In fact, mathematicians—who don't believe anything unless you give them a watertight proof—have no difficulty at all at defining something that nobody could ever see. They simply define higher-dimensional spheres in the same manner as they defined lines, circles, and balls: the collection of points in n-dimensional space (where n can be any number) that are not farther away from the center than the radius. Believe it or not, they can even tell you the volume of such an n-dimensional sphere (see the table in the appendix).

Let us return to packings and decide what we mean by its density. After all, we can always put an infinite number of spheres into an infinitely large space, so where does that leave us? Well, it leaves us with an example of why mathematicians are so nitpicky about seemingly obvious matters. So before we embark on any further investigation, the notion of density must be made precise. Mathematicians define the density of a packing as the ratio of the volume of the space that is filled by the spheres to the volume of the whole space. To compute the density, we must simply divide the volume that the spheres occupy by the volume of the space. This holds for any dimension and, in the limit, also for a space that extends to infinity. It may seem a wee bit difficult to measure the volume of an infinite space, but such minor impediments don't stop mathematicians. They define the density of a space as the limit of the above ratio as the space gets larger and larger.

Can you imagine what the densest packing of spheres is in one dimension? We already know that one-dimensional space consists of just a line, and that one-dimensional spheres are pieces of such a line—for example, matches or toothpicks. Now try to pack as many matches or toothpicks along a straight line as possible. It won't take you long to realize that the densest way to pack them is to place the matches end to end. In fact, this manner of packing achieves the best possible density: 100 percent of the line

---

[1] One can also define a curved line as a one-dimensional object. Then the spheres would be pieces of the curved line.

is filled with matches and there is no space left over in between. This is so obvious that even mathematicians do not require a proof.

Let us move to two dimensions. Here the problem is to place circles in a plane. We illustrate with a simple example. Take some coins of the same size, such as nickels, place them on a table, and push them around for a while. You will quickly find that the densest pattern is the one where each coin is surrounded by six others, that is, where the coins form a hexagonal pattern. You don't even have to be very careful when you place the coins, just push them around a bit and they usually arrange themselves into that pattern on their own.

What is the density of this pattern? Leaving the exact calculations for the appendix, we can see from the picture that the basic pattern that determines the packing is the hexagon, a regular six-cornered object. The whole surface can be regularly tiled with hexagons. Part of each hexagon is filled by circles, part of it stays empty. The hexagon can be partitioned into equilateral triangles, and each triangle is identical to the others. We can therefore restrict ourselves to computing the density of the triangles. As it turns out, the spheres cover 90.7 percent of the surface.

For comparison purposes, let us determine the density of the coins when they are arranged in a regular square packing. In this case the coins fill less than 79 percent of the surface (see the appendix for details of the computation). Hence, in two dimensions the regular square packing is much less efficient than the hexagonal packing.

It is important to note that the hexagonal packing is not necessarily a denser packing than the square packing unless the surface is extended to infinity. For example, using the hexagonal packing, we would be able to fit only three spheres into a square of edge-length four, while four spheres would fit into it when using the square packing. Something similar is also true in three dimensions, and Kepler's conjecture, the subject matter of this book, refers to space that has no borders, that is, that extends to infinity.

We saw that the hexagonal packing is denser than the square packing in

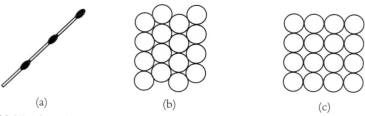

(a)   (b)   (c)

(a) Matches, (b) coins in a hexagonal packing, (c) coins in a square packing

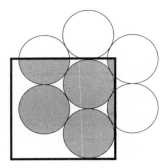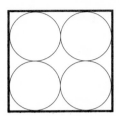

Packing in a finite box

two-dimensional space, but is it the densest possible packing? It is not at all obvious that no denser pattern exists, and the optimality of the hexagonal pattern does require proof. But even though the result looks quite banal, a rigorous proof was no simple feat, and it took until 1940 to find one that satisfied the community of mathematicians. We will return to that problem in chapter 4.

Back to Raleigh's cannonballs. Upon receipt of Harriot's letter, Kepler did not have to reflect for long in order to come to the conclusion that the densest way to pack three-dimensional spheres was to stack them in the same manner that market vendors stack their apples, oranges, and melons. In 1611 he published a little booklet that he presented as a New Year's gift to his friend Wacker von Wackerfels. It was called *The Six-Cornered Snowflake,* and in it he described a method of packing balls as tightly as possible. This marks the birth of Kepler's Conjecture. We will have more to say about snowflakes and their relationship to the packing of cannonballs in the next chapter.

Let us use melons as an illustration. If melons were cube–shaped, everything would be much simpler. They could be stacked side by side and on top of each other, with no space left over in between. As was the case with the matches, the density would be 100 percent. For exactly this reason attempts have been made to breed cubic melons.[2] Since produce is often flown from hot countries to overseas markets, melons have to be loaded onto airplanes. This could be done most efficiently with boxes loaded with cube-shaped melons. So why, the reader may ask, did nature evolve round melons (assuming, for illustration purposes, that melons are perfectly round

[2] Japanese farmers have figured out how to grow cubic watermelons.

objects)? And why are so many other fruits and vegetables approximately round? Well, nature did not worry about limited space on ships or in the holds of aircraft, but it did worry about moisture loss in hot countries. And it strove to minimize this loss. An object's loss of moisture is proportional to the object's surface: The more skin that is needed to cover the object, the higher the moisture loss due to evaporation. And which shape minimizes the surface for a vegetable of a given volume? As the reader may guess, the answer is the sphere.[3] If you compare two melons of the same weight, one cube-shaped and the other round, the round one has nearly 20 percent less surface than the cube-shaped one (see the appendix). By evolving round melons, nature strove to minimize the surface in order to reduce moisture loss. By the way, this is another of those vexing problems that took millennia to prove. Archimedes already knew the presumably correct shape. But only in 1894 did Hermann Amandus Schwarz (1843–1921) rigorously prove that the round sphere is the shape that minimizes the surface for a given volume.

Similar consideration may have led two mineral water distributors to dissimilar conclusions about the best shape of the containers they should use for distribution. One of the companies, by the name of Neviot, distributes water in cube-shaped canisters. The other, Eden, delivers cylindrical bottles. (Neither of them use round bottles, presumably because they would roll off the trucks.) Apparently, Neviot attempts to maximize the number of bottles it can fit onto a truck, and cube-shaped bottles do the trick. What does Eden do? They apparently try to minimize the cost of the raw material, since—for the same volume—cylindrical bottles require less plastic than the cube-shaped ones. But Eden containers do have an important advantage for the end user. The 20-kilogram bottles can be rolled from the front door to the kitchen, while Neviot bottles must be carried.

Returning to the fruit stand, one method of displaying the wares is to just place them helter-skelter into a box. With good reason, very few vendors choose this avenue. Not only is it a very unappealing way to show off melons, it is also inefficient. Experiments show that only about 55–60 percent of a box's volume is filled by randomly placed spheres. A better, though not much more esthetic, procedure is to shake the box while the melons are being poured in. Assuming that none of them are squashed in the process, about 64 percent of the container can be filled in this manner.

A more esthetic way to place the melons is to arrange the first layer in neat rows and columns, and then build the next layer by placing the next batch carefully on top of the lower melons. Obviously this cubic stacking

---

[3] This is one version of the so-called *Problem of Dido,* to which we will return in chapter 3.

method has a serious drawback: the melons are unstable. The slightest jolt from a customer would bring the whole stack tumbling down. But the stability of melon stacks—while of great concern to market vendors—carries no interest for mathematicians.[4] What does trouble them is that on an infinitely large table the cubic stacking method is inefficient. The density only reaches about 52 percent. So when the melon heap caves in, density actually increases by about 3–8 percent. Only dumb vendors would go to such lengths in order to build an unstable heap of melons that is also inefficient.

Shrewd vendors can do better than that. As it turns out, in markets all over the world the same universally accepted stacking method is used. First the individual fruits are placed along a line from one end of the table to the other. As we saw before, this is the densest packing in one dimension. Then the next line is filled in such a manner that each melon of the second line comes to lie not next to a melon of the first line, but next to the valley that is formed between two melons. In mathematical lingo, the second line is "transposed by half a melon." This goes on until the table is filled. Looking at the counter from above, the vendor now has the densest possible packing in two dimensions.

Let us go to the next layer of melons, which means moving into the third dimension. It is not quite obvious what the vendor should do. One method that we could devise would involve placing each melon of the second layer exactly above a melon of the lower layer. This results in a density of 60.5 percent (see the appendix). Unfortunately, this is not much better than the random arrangement of melons.

But shrewd mathematicians can do even better than that. They are quick to point out that between every three melons of the first layer a dimple has formed. A larger quantity of fruit can be stacked if the melons of the second layer are placed into the dimples of the first layer. On the next layer one dimple is filled with a melon, the next dimple is left empty, one is filled, one is left empty. And so on. As we will see in chapter 2 the density of this so-called *hexagonal close packing* (HCP) reaches a whopping 74.05 percent. Not only is this way of stacking melons better than the previous one, it is the *best* way to stack melons. In other words, it is the densest packing. Market vendors know it, you and I know it, and Harriot and Kepler knew it, but mathematicians refused to believe it. And it took 387 years to convince them of the truth of this fact.

At this point I want to divulge two interesting and very important facts about the packing of spheres. They indicate that nothing is as simple as it

---

[4] Physicists, on the other hand, do worry. See what Per Bak has to say about the stability of *sandpiles* in *How Nature Works* (New York: Copernicus, 1996).

looks, especially in mathematics. In 1883 the crystallographer William Barlow (1845–1934) pointed out that there is not just *one* good way to stack melons, but *two*. Barlow was a self-educated scientist who used the leisure afforded by an inheritance from his father to study and work in crystallography. Convinced that the manner in which atoms and molecules are packed around each other provides the answer to the symmetrical forms of crystals, he investigated different packing arrangements. After many years of study, he published an article in the British journal *Nature,* in which he described five arrangements of atoms in space. Two of them are of interest here.

The first arrangement is the market vendor's HCP packing that was described previously. But let us inspect the strategy of the dumb vendors again for a moment. They start out by placing their melons in neat rows and columns. Haven't we already rejected that arrangement as being inefficient? Well, the crucial point is the next layer. Note that there are dimples again, but this time they exist between every four melons. (In the HCP packing, there are dimples between every three melons.) The dumb vendors place the melons of the next layer into these interstices, and start building up the stack. They receive a packing called the *face-centered cubic* packing (FCC). Why would vendors do such a dumb thing, after we have shown that the HCP is the most efficient arrangement? Well, the HCP is the most efficient stacking method, but it is not the *only* one. Upon close—very close—inspection, it turns out that the FCC and the HCP are the exact same packing, viewed from different angles! This seems rather unbelievable at first. But a very instructive illustration in Barlow's paper, which depicts a cutaway of an FCC arrangement, proves that both arrangements are, in fact,

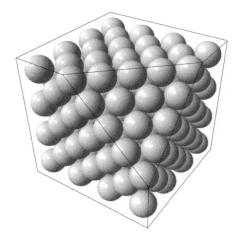

Barlow's picture

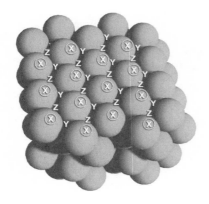

There is an infinite number of ways to stack spheres

equivalent. Since the FCC and the HCP represent the same packing, they must, of course, have the same density of 74.05 percent. So the dumb vendors aren't so dumb after all.

Twenty-four years later the amateur scientist struck again. Together with his colleague William Jackson Pope, later professor of chemistry in Manchester, Barlow wrote a paper that appeared in the *Journal of the Chemical Society* in 1907. In this paper the two men showed that there are not just two, but an *infinite* number of ways to stack melons in the most efficient manner. (Actually, they were concerned more with atoms than with melons.) Let's describe what they meant, by using the HCP.

After having arranged the first layer of melons, the vendor must make a decision: Which dimples should he use for the second layer? He could use the interstices marked with a Y in the picture. Or he could use the ones marked with a Z. Let's say he uses Y. In the following layer he again faces a choice: should he use the interstices marked Z or those marked X? And so on. After a few layers, the heap is stacked as XYZXZX . . . , or as XZXZYX . . . , or as XYXYXY . . . , or as any other succession of layers from among an infinite variety of possibilities. All of these arrangements have a density of 74.05 percent! Wouldn't Harriot and Kepler have been surprised?

Do the infinitely many packings have something in common, apart from their density? Yes, they do. In every one of those arrangements each sphere is in contact with twelve others. But don't confuse this statement with its converse. Not every arrangement in which a sphere touches twelve others is efficient. In fact, there exist very obnoxious arrangements that I will call the *dirty dozens.* I will have more to say about them in later chapters. For the time being, let's just agree that, definitely, nothing is as simple as it looks in mathematics.

CHAPTER 2

The Puzzle of the Dozen Spheres

Harriot's pen pal, Johannes Kepler, was born near Stuttgart, Germany, to Heinrich and Katharina Kepler (née Guldenmann). Heinrich and Katharina got married on May 15, 1571, and seven months later, on December 27 of the same year, little Johannes was born. Lest one believe—perish the thought—that Katharina was already pregnant on her wedding day, note that Johannes was born prematurely. He himself maintained that he had been conceived on the morning after the wedding night, at thirty-seven minutes past four o'clock. The precise moment of conception was of some importance to Kepler since this learned man, the foremost scientist of his time, dabbled in astrology from time to time.

His parents did not provide what one would call a warm home. His father was an extremely unlikable man, described by contemporaries as a bad-tempered hothead, and his mother was not much better. A small, thin woman, garrulous and quarrelsome, she was known to have an exceptionally vile character, and apparently devoted her existence to making life miserable for Heinrich. Annoyed, he fled his home to join the Spanish army, leaving 3-year-old Johannes at home with his mother. But Katharina, not one to accept defeat, set out in search of her husband. She finally caught up with him in Belgium, and we can only imagine Heinrich's embarassment in front of his warrior companions when this woman suddenly appeared out of the blue. Left with no choice, Heinrich followed her back to Germany. But he could not bear the homestead for long; he yearned for his drinking and brawling days. Soon he snuck off again to rejoin his war buddies. While living it up once again in Belgium, he committed some unknown misdemeanor and only narrowly escaped the gallows. Three years later he returned to Germany, in ill health and looking much worse for wear, to try his hand as an innkeeper. But that career change did not suit Heinrich either and, fed up with his wife's constant bickering, he eventually decided that enough was enough. One day he left the house, never to be seen again. How and where his life ended is not known.

In such an unfavorable environment, the as-yet-hidden gifts of the young Johannes would never have stood a chance of manifesting themselves, had it not been for a gifted children's program that the local noblemen, the dukes of Württemberg, established in the town of Leonberg. Johannes was accepted to the school. He excelled in his studies but did not become the life of the party. Apparently he had inherited his mother's disagreeable disposition, was nasty to most of his classmates, and constantly got involved in fights and petty arguments.

When Kepler eventually wrote his life story, it read, in part, like this:

> Holp openly detested me and on two occasions we got into fist fights. . . . Molitor disliked me because I had betrayed him and Wieland. . . . Köllin didn't hate me, but I hated him. . . . Braunbaum turned from friend to foe because of my boisterousness. . . . Huldenreich became hostile because of my rash accusations. . . . Seifert I disliked because everyone else disliked him. . . . Ortholf could not stand me. . . . Spangenberg was angry at me because I corrected him, even though he was the teacher. . . . Kleber detested me because he thought I was a rival. . . . Rebstock was ticked off whenever someone praised my abilities. . . . Husel tried to block my progress. . . . Between Dauber and myself there was a quiet jealousy. . . . Lorhard would have nothing to do with me. . . . After Jaeger had lied to me I was insulted for two years. . . . The rector became my enemy because I did not accord him sufficient honor. . . . Murr became my enemy because I reprimanded him.

And so on, and so on. Not once did Kepler mention a friend, except to say that one had also turned into an enemy. Of course, the bad vibes weren't the poor boy's fault. The deeper reason for all this hatred and resentment was that, as Kepler himself put it, "Mercury was in the square of Mars, the Moon in the trine of Mars and the Sun in the sextile of Saturn." On top of that Kepler was a hypochondriac, suffering from one illness or another throughout his youth, although there was no astrological explanation for that.

But he did manage to learn Latin at school. This would come in handy later on, since it was the *lingua franca* of science at the time, much as English is today. After three years of study, Kepler successfully passed the state exam and obtained one of the coveted places at the convent schools of Adelberg (where the day started at four o'clock with the singing of psalms) and, later, at Maulbronn. In 1589, half a year after his father's final departure, the newly graduated *Baccalaureus* entered college with the idea of eventually

Johannes Kepler

becoming a man of the cloth. But, as was the custom at the time, Kepler had to take two years of classes in the faculty of arts of Tübingen University before embarking on the study of theology. After receiving the Master of Arts degree, Kepler was finally allowed to enter the theological faculty. The year 1594 saw him near the end of his studies, and Kepler started looking around for a job as a clergyman. But to his great chagrin a hitch developed that would prove of everlasting benefit to science and to the world.

One of his teachers, the professor of mathematics and astronomy Michael Mästlin, had noticed in this young prodigy an extraordinary talent for science. He therefore recommended that upon graduation Kepler be sent to the Austrian town of Graz to serve as mathematics teacher in the cathedral school. The theological faculty of Tübingen was also not unhappy to rid Kepler of his ecclesiastical ambitions, since he had shown too independent a mind for their taste. The problem, in their eyes, was that he showed an interest, fueled by his teacher Mästlin, for the Copernican solar system. That system had the sun in the center of the universe instead of the earth and was, therefore, frowned upon by the holy men. So, in spite of his protestations, Kepler was sent to Graz to begin his duties as a schoolmaster.

His new location turned out not to be quite as bad as he had feared. He set eyes on a young noblewoman, Barbara Müller von Mühlegg, and decided to seek her hand in marriage. But before he could wed his heartthrob, her family insisted that the bridegroom prove that he was of noble descent. Kepler journeyed back to his hometown to procure the required documents. He succeeded in obtaining them, but upon his return to Graz after several months he found that some rivals had almost convinced Barbara to forget about him. It took no little effort on his part to change her mind again. The wedding finally took place on February 9, 1597.

Teaching did not fulfill him, and Kepler kept himself busy reworking the town's calendar. This task did not just involve the assignment of weekdays to the days of the months, but included astrological predictions. Kepler foretold a few political events—based more on common sense than on the position of the planets—which turned out to be true. He also predicted a freezing winter, and according to contemporary accounts the season actually turned out to be so icy cold that people's noses fell off when they blew them! Such prophesies increased his stature immensely among the townsfolk, though not among the faculty and senate of his alma mater. In spite of his being a devout Protestant, Kepler had used the calendar reform introduced by Pope Gregory XIII in 1582. The Lutheran senate of the University of Tübingen did not hide its displeasure.

This venture rekindled Kepler's interest in astronomy—in the number of planets, their sizes, and their orbits. His religious beliefs remained firm, however, and Kepler sought a theological explanation to his questions. Since God had created a perfect world, he thought it should be possible to discover and understand the geometric principles that govern the universe. After much deliberation Kepler believed that he had found God's principles in the regular solids. The key idea, so it is said, came to him in the middle of one of his classes. His explanations of the universe were based on an imaginary system of cubes, spheres, and other solids that he thought were fitted between the sun and the planets. Kepler wrote up his theory and published it in a book entitled *Mysterium Cosmographicum*. This tome did not unveil any mysteries of the planetary system. It couldn't have, since no solids exist that are suspended in the universe. But the book did come to the attention of Tycho Brahe, the great Danish astronomer.

Brahe was born in 1546, the first son of a noble Danish family. Problems started before his birth because the father had promised his brother, a childless vice-admiral, that if the newborn would be a boy, he would let him adopt and raise him. But when the father first set eyes on the cute little baby, he reneged. Uncle Jörgen accepted this with understanding, but after a second son was born to Mr. and Mrs. Brahe, Jörgen thought that surely they had no further need for their firstborn and without much ado kidnapped Tycho. The father thought otherwise and threatened to kill his brother. He only calmed down when he realized that his son stood to inherit a great fortune from the childless uncle. The young boy was sent to study Latin so that he could eventually become a lawyer and enter Denmark's civil service. But at age thirteen, Tycho witnessed an event that would shape his career: he observed a partial eclipse of the sun that had been predicted for that day. The openmouthed boy decided then and there that astronomy would be his profession. But first he had to begin his law studies in the German town of Leipzig. His secret infatuation with astronomy never waned, however.

Brahe moved to Augsburg in Germany and joined the local stargazer's club. The young man convinced his newly found amateur astronomer friends that what they needed were more accurate observations, so the club commissioned a wooden sextant—a contraption with which a skilled astronomer could determine the position of the stars—with a diameter of 12 meters. Along its edge notches were cut approximately every 1.5 millimeters, which corresponds to gradations of 1′ (1′ is one-sixtieth of a degree; 360 degrees form a circle). This allowed the determination of the whereabouts of celestial objects to an unprecedented accuracy.

His rich-kid upbringing made Brahe somewhat of a spoiled brat, and he liked to think of himself as the best and the brightest. One day he got into an argument with another student about who was the better mathematician and it was decided to settle the question once and for all in good Teutonic fashion: by fighting a duel. In the course of this contest, part of Brahe's nose got cut off. It is not recorded what body part, if any, his opponent lost and so conclusive proof as to who was the better mathematician is missing.

It is certain, however, that Brahe was a gifted inventor of astronomical instruments and that he was equally talented in using his equipment. He must also have had a high tolerance for boredom since he was able to sit for hours in an observatory looking at the stars. King Frederick II of Denmark, a patron of the arts and the sciences, made Brahe an unprecedented offer: the picturesque island of Hven would be his, together with a castle and all the amenities that a scientist could wish for. The property came with a paper mill and a printing press, and all inhabitants of the island were to be Brahe's subjects. The castle even had its own little prisonette, in which disobedient farmers could be incarcerated. Brahe graciously accepted and set out to build a magnificent observatory, which he called Uraniborg.

Even though the master liked to entertain visitors, and many evenings were filled with parties and festivities, Brahe spent most nights during the next twenty years sitting in his observatory with his assistants, following and recording planetary motions. At times, four teams of observers and timekeepers measured the same thing simultaneously, thus reducing errors to a minimum. Brahe performed his measurements not only to an unparalleled precision but also, and equally important, with uninterrupted continuity. His meticulously kept journals were to be the key to a new understanding of astronomy. And did he ever know it. He jealously guarded them like a treasure and did not permit anyone access to their content. The celestial system he envisioned was to become an improvement on both the Ptolemaic system, which placed the earth at the center of the universe, and the Copernican system, which put the sun at the center of the planets' circular orbits. Brahe proposed a system—which was, of course, to carry his name—that put the earth at rest, had the sun going around the earth, and all planets going around the sun.

But before he was able to achieve his aim, the haughty scientist picked a quarrel with the Danish king. Apparently fame had gone to Brahe's head and he had become quite a tyrant toward the inhabitants of the island Hven. The prisonette that came with the castle became instrumental in his downfall. Brahe took his disciplining authority seriously and had an unruly farmer and his family imprisoned. The poor man appealed to the High Court of Denmark and the judges, in a remarkable display of equality before the law, ordered Brahe to free the man. But the cocksure astronomer would have nothing of that, and kept the man in chains. By that time, however, King Christian IV had ascended the throne, and the new king was no longer as well disposed towards the vain and conceited astronomer as had been his predecessor. Fed up, the king reduced the salary that came with Brahe's cushy job. This didn't sit well with Brahe, and the deeply insulted astronomer packed up his instruments, took his family and his journals, and left Denmark.

It took Brahe two years to land a new job, but in 1599 he got lucky. And what a job it was. Brahe was asked by Emperor Rudolph II to become "Imperial Mathematician" to the court in Prague. Among the many perks of the job was the prospect of a position—although no money—for an assistant, and Brahe immediately remembered Kepler. Actually, at the same time the young teacher was about to embark on a job hunt himself. Problems had developed between his Protestant school and the Catholic town of Graz. All members of the faculty were forced to sign an affidavit about their religious denomination. Kepler, unwilling to lie, declared that he was a Protestant, knowing full well that his convictions would sooner or later get him booted out of Graz. Five days after his twenty-ninth birthday, on January 1, 1600, Kepler left the town for a half-year stint in Prague. He returned the following June, but only to pack up. In September he left for good with his wife Barbara and his child, to take on the post of Mathematical Assistant to the Imperial Mathematician. His salary was to be paid by his new master.

The collaboration between the master and his new assistant did not turn out to be an easy one. Brahe gave Kepler the task of figuring out the movements of the planets, which, as he had already discerned, did not follow circular orbits. The assistant was supposed to do this based on Brahe's own observations of the stellar positions, but without getting full access to the data. Only when the master felt like it, did he hand out bits and pieces of the compiled data. Apparently, he was afraid that the bright assistant would surpass him. And he was quite right to fear his devious assistant. "Tycho is very stingy as to communicating his observations. But I am allowed to use them daily," Kepler wrote to his former teacher, adding, "if only I could copy them quickly enough!"

To keep him busy, Brahe put the frustrated Kepler to work on his observations of the planet Mars, which had the least circular orbit. Kepler soon discovered that the orbit of Mars is an ellipse. Fortunately for Kepler— unfortunately for Brahe—the fruitless relationship ended after one year, when the latter died from a bladder infection. The illness had afflicted Brahe after he had indulged in an especially sumptuous meal. Emperor Rudolph II lost no time in promoting Kepler to the post of Imperial Mathematician, and the erstwhile assistant inherited Brahe's prized possession, the journals. Actually, *inherited* may be too nice a word. *Stole* would more aptly describe the act with which Kepler managed to take hold of the log books before Brahe's heirs could lay their hands on them.

Unfortunately the lofty title that went with the imperial job did not carry with it a commensurate financial reward, since the imperial treasury was close to empty. Nevertheless, Kepler labored day and night on an explanation of the planetary system and was finally able to publish his *Astronomia nova* in 1609. But the hard work had taken its toll, and Kepler soon suffered various illnesses and bouts of depression. In 1612 his wife and his favorite son died of fever and smallpox. Struck with sorrow about his son's death— he did not care much about his wife even while she was alive—Kepler wanted to leave Prague, which had become a cauldron of religious and political strife. But he was forced to remain, and only when his sponsor, Emperor Rudolph II, died a few months later was he allowed to depart.

Kepler moved back to Austria, made his home in the town of Linz, and took another wife, Susanna Reuttinger. For the next fourteen years he found sufficient calm in Linz to continue his work on astronomical tables and publish some of his fundamental works. But life was not all roses. Religious doubts tormented Kepler, and when he confided in a clergyman, the good man lost no time in making Kepler's hesitations known to the proper authorities. The astronomer was told to lay off theological speculations and concentrate on mathematics.

Then, out of the blue, another catastrophe hit: Kepler's mother, Katharina, was accused of being a witch. Knowing her disagreeable temperament, accusations of ill temper would not have been surprising, but a witch? That seemed a bit too much, even for a woman of Katharina's exceptionally obnoxious character, since a conviction carried the death sentence. Katharina's own aunt had suffered that fate a few years previously. What had actually happened was that Katharina, who liked to collect all kinds of herbs with supposedly medicinal powers, had got into a dispute with some old hag, a certain Mrs. Reinhold, who had previously been her best friend. This woman, Katharina's match in nastiness, decided to pay her back by accusing Katharina of having given her a potion that made her suffer bouts of depression. All of a sudden other townsfolk remembered that they too had fallen ill after drinking Katharina's potions, and the witch hunt started.

Kepler spent six difficult years fighting for his mother's innocence. Katharina was arrested by the authorities and Kepler found her in the prison of Güglingen, bound hand and foot and guarded day and night by two watchmen. Since Kepler had to pay the guards' salaries, he wrote a letter to the court asking whether one watchman would not suffice to guard this seventy-three-year-old woman, all the more so since she was shackled to a wall anyway.

Against all odds, Kepler managed to get a verdict of innocence for his mother. Actually, Katharina herself was quite instrumental in winning her eventual release. The generally used interrogation method of the time was to extract a confession by torture. In Katharina's case the court made an exception. It was decided to lead the suspected witch to the torture chamber and show her the instruments with which she would be tormented if she didn't tell the interrogator what he wanted to hear. The good men believed that the mere exhibition of the tweezers, pulleys, chains, red-hot irons, and other equipment would make her admit to any wrongdoings. Any lesser woman would have succumbed to such gentle persuasion, but not this stubborn lady. She would not confess. The increasingly frustrated torturer hauled over more and more sinister looking instruments, but to no avail. Finally the old woman fainted and the poor torturer gave up. Based on such conclusive evidence, the court declared Katharina Kepler innocent in October 1621. But the old woman was not to enjoy her victory for very long; half a year later she died.

Kepler managed to complete work on the astronomical tables and then traveled on to Prague where he presented his work, duly dedicated to his excellency, at court. The emperor was well pleased and awarded Kepler an extraordinary honorarium of 4,000 florins. This generous gesture cost him nothing and did Kepler little good, since it was simply added to the sum the emperor already owed the imperial mathematician.

Desperately looking for funds, Kepler rode 500 kilometers on horseback to the town of Regensburg. There he hoped to collect the emperor's outstanding debt from the Reichstag, the assembly of German dignitaries, who should have honored the emperor's promises. The trip took nearly four weeks. Foul autumn weather got the better of Kepler. Ill and destitute, he eventually arrived in Regensburg to find shelter at a friend's house. But his last respite was to last barely a week. On November 15, 1630, six weeks before his fifty-ninth birthday, Kepler, the prince of astronomy, died.

Kepler's legacy included many significant achievements. His main triumph, however, was to bring the Copernican system into its final form. As we noted, Copernicus believed that the planets of the solar system turned in circles around the sun. Kepler took a closer look at the meticulously compiled observational data he had inherited—or pilfered, depending on whose side you are on—and realized that the planets actually move in ellipses, with the sun at one of the focal points. This was the so-called first law of planetary motion. He was able to formulate two more laws: The speeds of the planets' movements are not uniform, and the squares of the periods of evolution of any two planets are proportional to the cubes of their distances from the sun. He arrived at these results simply by gazing at the series of numbers—as he gazed at the planets themselves—until a pattern appeared among the numbers.

But Kepler did not only concern himself with big questions about the heavenly bodies, he also showed an interest in the small-scale workings of nature. And here lies his importance for our investigation. Kepler's conjecture is contained in the little booklet *The Six-Cornered Snowflake,* which he wrote in 1611 as a New Year's gift to his friend Wacker von Wackerfels. Von Wackerfels, a traveler, intellectual, and sometime diplomat for the Bishop of Breslau (Wroclav), had become friendly with Kepler during his stay in Prague.

Written in a colloquial style in Latin, the booklet is rife with private jokes that few people now understand, and contains numerous references to Greek mythology and natural sciences that only well-read contemporaries would grasp. Apparently, both Kepler and von Wackerfels kept abreast of the current list of bestselling authors—Virgil, Homer, Aristophanes, Euripides—and knew all the references. (Well, not quite. In his efforts to appear well versed, Kepler refers to a fable by Aesop that simply doesn't exist.)

Compared with the importance of his voluminous tomes on astronomy, the little treatise on snowflakes made few waves, and it is hardly mentioned in some of the bibliographies of Kepler's works. But its importance must not be overlooked. This was one of the first times ever that an attempt was made to explain the physical forms of crystals and plants with the tools of a scientist. Kepler's contemplations in this little booklet have been described by a biographer as "a scientific judgment of the highest caliber."

In the treatise, Kepler investigates the reasons why snowflakes are shaped as they are. But before turning his attention to snow, Kepler examined three questions concerning nature's forms: why honeycombs are formed as hexagons, why the seeds of pomegranates are shaped as dodecahedra, and why the petals of flowers are most often grouped in fives.

Bees build their hives in a pattern of hexagons that are closed at the bottom by three quadrangles. Hence all bees, except those in the "corner offices," are surrounded by six neighbors on the sides, and by three more neighbors on the bottom. The little creatures leave the tops of the hives open, for if they added roofs—so the imperial mathematician astutely pointed out—they would have no exit from their little condominiums. But it was the floorplan that caught Kepler's attention, and he wondered about possible explanations for its hexagonal shape. He thought that there were three reasons. His step-by-step deductions are beautifully reasoned, even though there are some holes in the arguments.

First Kepler observed that the hexagon is a shape that tiles a surface without gaps. But then so do squares and triangles. So why hexagons? The answer, according to Kepler, is that the two other shapes have less surface than the hexagon on which honey can be stored. Here the great scholar resorts to a bit of handwaving, since he neglects to specify a common scale with which the surfaces of the three geometric forms could be compared in a fair manner. After all, a large triangle has a bigger surface than a small hexagon. What he probably meant was that when comparing squares, triangles, and hexagons with equal edge sizes, the hexagon wins. Or that among triangles, squares, and hexagons with the same surface, the latter minimizes the lengths of the walls.[1]

Be that as it may, the master then turns to a second reason for the honeycomb's hexagonal form. The most comfortable dwelling for bees is, according to Kepler, circular. And of the three shapes that can tile a surface, the hexagon's lack of sharp corners makes it resemble a circle most closely.[2] Therefore, it is superior to triangles and squares. But if comfort, storage space, or the parsimony of building materials were the guiding principle for the bees' condominium, than why not build round hives? After all, less wax is required for the construction of the walls of a round room than for the walls of a hexagonal room with the same surface. Here comes the final reason for the hexagonal shape of the bee dwellings, which is itself threefold.

---

[1] This is also a version of Dido's problem (see chapter 3).
[2] In 36 B.C. the Roman scholar Marcus Terentius Varro gave a very plausible explanation: he thought that beehives were hexagonal in order to accommodate the bee's six legs.

First, in the construct-your-own-hexagon project, neighboring bees can share the work by building common walls, one of them working on either side. In a build-your-own-circle scheme each bee would be on its own. Next, Kepler claims that the straight edges of the hexagon give the hive more stability than do round walls, and make it less susceptible to crushing.[3] Finally, and maybe most convincingly, gaps between the spheres would let the cold seep into the hive and should, therefore, be avoided. Actually, Kepler here resorts back to his initial argument, since gap-less floorplans were the requirement he started out with. All these reasons should suffice—so Kepler believed—to explain why the Creator imprinted bees with the archetype of the hexagonal pattern. He saw no need to delve into further arguments, but did think it appropriate to also mention "the beauty, perfection, and noble figure" of hexagons.

Kepler then turned his attention to pomegranates and it is here that he formulated the conjecture that was to keep mathematicians, imperial and otherwise, busy for nearly four centuries. Kepler observed that the seeds of pomegranates are of a rhomboid shape, with twelve faces. He correctly surmised that this shape arises when the seeds are pressed together in the constrained space of the fruit. As long as they are small, the seeds are round and float around freely. But as they grow, they apparently organize themselves in such a manner that each seed in the crowd is surrounded by twelve others that are pressed against it.

This observation was ostensibly confirmed more than a century later, in 1727, by the English botanist Stephen Hales (1677–1761).[4] In his book *Vegetable Staticks,* the world's first work on plant physiology, Hales reported experiments on plant respiration and transpiration. In one of these experiments he put peas into a pot and then applied pressure. "I compressed several fresh parcels of Pease in the same Pot, with a force equal to 1600, 800, and 400 pounds, in which Experiments, tho' the Pease dilated, yet they did not raise the lever, because what they increased in bulk was, by the great incumbent of weight, pressed into the interstices of the Pease, which they adequately filled up, being thereby formed into pretty regular Dodec-

---

[3] Engineers would dispute this stability argument. Even in the Middle Ages church builders knew, for example, that arches above doorways could support more weight than flat doorframes.

[4] Stephen Hales's namesake, Thomas Hales, the mathematician whose struggles with Kepler's conjecture will be described in the later chapters, wrote of this botanist: "Even though Stephen Hales married late and died without children, so that we can't claim him as an ancestor, the scientists in my family have informally 'adopted' him into our family."

ahedrons." Had he used a force of more than just 1,600 pounds, his vegetables would have been pulped into pea soup, instead of emerging as regular, twelve-sided dodecahedra.

Or so Hales thought. Unfortunately, his conclusion was incorrect. After all, it is not so easy to tell whether a pea has been squashed into a dodecahedron or into some other kind of polyhedron. Presumably Hales was led to believe that he saw dodecahedra by observing many pentagonal faces on the peas. (A regular dodecahedron has pentagonal faces.) But they could not all have been dodecahedra because of one irrefutable fact: in the same manner that pentagons cannot tile a floor, dodecahedra can't fill three-dimensional space without gaps.

I will have more to say about that in later chapters, but before we leave Hales, we must partially rehabilitate him. He did not really claim that squashed peas became regular dodecahedra. Rather, he said that they become *pretty regular* dodecahedra. While he may have found dodecahedra pretty to look at, it is more likely that he meant that the dodecahedra he saw weren't exactly regular. Hales's experiments were repeated in the 1950s by various physicists, such as J. D. Bernal and G. D. Scott. They compressed ball bearings in various packings and determined that the FCC, the face centered cubic packing of chapter 1, produced the tightest packing.

Hales's somewhat faulty conclusion did confirm Kepler's observation, however. After seeds and peas have been squashed and squeezed, they will all be touched by twelve neighbors. If seeds accidentally found themselves placed along parallel rows and columns, and seeds of the next higher planes came to lie exactly above the ones of the lower planes, they would become cubes when pressed together. But—and here Kepler performs an enormous leap of faith—this does not represent the densest packing: *"sed non erit arctissima coaptatio."*

There are a couple of problems with this statement. First, Kepler takes it for granted, without further justification, that nature, or the Creator, arranges the seeds in the tightest manner possible. Second, he claims—also without justification—that the tightest packing is achieved when twelve seeds arrange themselves around a central seed in a regular manner. That is why, so he claimed without further ado, that when they are pressed together, the central seed is deformed into a twelve-faced rhomboid, and not a dodecahedron, as Hales had thought. So, here it is. On pages 9 and 10 of Kepler's little treatise we find the formulation of his famed conjecture: A sphere surrounded in a certain way by twelve other spheres represents the tightest possible packing. Kepler didn't prove this statement, he just announced it.

Undeterred by such imperfections, Kepler then went on to describe a remarkable fact. Such a packing can be built up from two different arrangements, one with a quadratic, or square, base and another with a hexagonal base. He even illustrated the contention with pictures. I will now refer to spheres instead of pomegranate seeds. If the basic arrangement is quadratic, all spheres lie along horizontal rows and vertical columns. The balls of the next layer come to lie in the crevices that are formed between four spheres of the first layer. The same holds for the next layer, and for the one above that one, and so on. In such an arrangement each sphere is touched by four spheres of its own plane—one each to the north, south, west, and east—by four spheres from the lower plane and by four spheres from the upper plane. We recognize the FCC.

Now on to the hexagonal base. Here six spheres in a plane surround the central sphere. In between these seven spheres there remain six holes—or interstices. Spheres are placed into every second interstice to form the next layer. The same is done in the layer below. And so, again, a total of twelve spheres touch the central sphere: six in the plane, three above, and three below. We now have the HCP, the hexagonal close packing.

Now Kepler points out something truly remarkable: the two arrangements are equivalent. That is, the two packing methods are exactly the same, the only difference being that they are perceived from different angles. This fact, which was rediscovered by William Barlow in 1883, implies, of course, that the two arrangements have equal density. On the one hand this may seem plausible, because in both cases twelve neighbors touch the central sphere.[5] On the other hand, the reader may find it strange that both arrangements have the same density. After all, in the square arrangement one sphere is placed in the dimple that is formed by four spheres. In the hexagonal arrangement, however, an additional sphere is placed in the dimple between three spheres. One may be led to believe that the latter is denser (one additional sphere for every *three* spheres, versus one additional sphere for every *four* spheres in the square arrangement). But this is not so: the interstices in the square arrangement are deeper than the interstices in the hexagonal arrangement. As it turns out, for the computation of density, the number of spheres that form the dimple and the depths of the dimples exactly offset each other. It is not easy to discern the equivalence of the two arrangements, and it takes some imagination to convince oneself that the two packings are, in fact, identical. The following picture, one hopes, makes this clear.

[5] However, we shall see later that there are more ways to have a dozen spheres touch the central sphere, and they may have different densities. The discussion between Isaac Newton and David Gregory (see chapter 5) concerns exactly this point, as does the "dirty dozen" (see chapter 11).

(a)   (b)

These two arrangements are equivalent

There remains Kepler's question of why flower petals are grouped in arrangements of fives, or why, for example, the seeds in apples and pears are housed in five cells. Contemplating this phenomenon, Kepler was led to speculate about the beauty of the number five. Since this number is associated with plants and fruits, he concluded that it must have something to do with life. In the deduction process he made some dramatic leaps of faith: first, he started out with regular geometrical bodies, which in his world-view are the basic building blocks of the universe. Second, one of those bodies, the dodecahedron, is made up of five-sided regular figures, pentagons. Third, the construction of a pentagon requires the so-called divine proportion (more on that follows), which can be derived as the outcome of a series of numbers. Fourth, a series of numbers, where preceding terms give rise to the succeeding ones, is a symbol of fertility. And, fifth, from fertility it is just a short step back to the seeds of apples and pears. Hence, the master concluded, the number five must have special powers as a symbol of fertility, and this "explains" why it should appear in flowers and fruits.

Let us examine Kepler's thought process in a little more detail. The number five appears in two regular bodies, the dodecahedron and the icosahedron. The dodecahedron's faces are pentagons, and five faces of an icosahedron form a pentagon. Drafting a pentagon requires a ratio that is called the *divine proportion,* or the *golden number.* This divine proportion can be calculated from a very special series of numbers. A series of numbers consists of a few terms, which are specified, and a rule about how the following terms are spawned from the previous numbers. The famous Fibonacci series, discovered by Leonardo of Pisa at the beginning of the thirteenth century,[6] specifies the first two terms as ones, and then defines the other terms in the series as the sum of the two previous numbers:

1, 1, 2, 3, 5, 8, 13, 21, 34, 55, 89, 144, 233, . . .

For example, 5 plus 8 equals 13, 8 plus 13 equals 21, 13 plus 21 equals 34, and so on. Now compute the proportions between pairs of consecutive numbers:

$\frac{1}{1} = 1$, $\frac{1}{2} = 0.5$, $\frac{2}{3} = 0.666$, . . . , $\frac{144}{233} = 0.618025$ . . . , . . .

As we move to higher terms, the ratio approaches a limit: 0.6180339. . . . This number, as well as its inverse, $\frac{1}{0.618033}$ . . . = 1.6180339 . . . , are called the golden numbers. Anything designed by nature or by man according to the ratio of approximately 0.618 to 1, or of 1 to 1.618, was thought to be especially pleasing aesthetically. That is why the ratio is called the divine proportion. It often appears in geometry and is omnipresent in nature. Artists and craftspeople try to incorporate the ratio in their artwork, and many buildings, paintings, and statues are based on the golden proportion.

We now see how Kepler got from the pentagon to the divine proportion, but how did he get from the divine proportion to fertility? That numbers in a series give "birth" to new numbers may, if you stretch your imagination, be an indication of fertility. But there is another explanation. Fibonacci discovered the series that was to carry his nickname during an investigation of how rabbits multiply. He started with the assumption that a pair of rabbits reaches maturity after one month, and then has a pair of offspring every month after that. Let us say that in January there is just one pair of newborn rabbits. In February this pair matures, and in March it has its first pair of offspring. There are now two pairs. In April another pair of offspring are born to the original pair, while the first offspring are still maturing. Altogether there are now three pairs. In May, the first pair and their offspring will each have offspring, while the April offspring are still maturing, for a total of five pairs. Come June, the first pair and their offspring will each have offspring, and the March offspring will also have kids, but the April offspring are still maturing. Altogether there are now eight pairs. At that point things become a wee bit complicated.

So the population of rabbits grows according to the Fibonacci series, and grows and grows and grows. But Leonardo of Pisa forgot to take into

---

[6] The son of Bonacci—therefore the nickname Fibonacci: *filius Bonacci.*

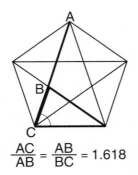

$$\frac{AC}{AB} = \frac{AB}{BC} = 1.618$$

Divine proportion in the pentagon (top), pentagons in the dodecahedron (bottom left), and icosahedron (bottom right)

account a minor detail: Rabbits have the funny habit of eventually dying.[7] So the population does not quite grow out of bounds, as the series suggests. Nevertheless Fibonacci's somewhat faulty reflections on the rabbits' reproductive behavior convinced Kepler that this series symbolizes fertility, and from fertility one is quickly led back to the seeds of apples and pears. That is why, according to Kepler, plants, fruits, and vegetables have a partiality to the number five.

Kepler's farfetched chain of arguments has no basis in reality, of course, but as is believed today, this devotee of astrology and esoteric symbolism may not have been far from the truth. For example, the seeds in a sunflower and the cells of pinecones and pineapple are arranged according to consecutive numbers taken from the Fibonacci series. This also seems to be true, in general, of the arrangement of branches around the trunk of a tree or of leaves around the stem of a plant. And so it appears that phyllotaxis (the science of how the physical form of plants evolve) is at least partially based on Fibonacci's celebrated series. Why this should be so is a mystery even today.

---

[7] And the more there are, the faster they die, because of an increasing scarcity of resources.

Having done away with bees, pomegranates, apples, and pears, Kepler was ready to attack his original project and get down to the nitty-gritty of snowflakes. The first question was, Why are these objects flat rather than three-dimensional? After some deliberations Kepler came to the following conclusion: snowflakes must originate when a warm front of air hits a cold front, and since this can only occur in a plane, snowflakes must necessarily be flat. The explanation seems quite plausible but there is one problem: it's not true. The size of the interface between hot and cold air is large, and both two- or three-dimensional snowflakes could form there as long as their diameters are small. The true reason for flatness will be discussed later.

Then Kepler turned his attention to the six-corneredness. He suggested that the snowflake's visible features may be caused by properties of the building blocks of the flakes, that are themselves so small that they cannot even be seen. This remark is very significant, because it was one of the first times since the Greek thinker Democritus had postulated the existence of atoms in the fourth century B.C. that a scientist proposed an atomistic explanation to a natural phenomenon. Actually, it was Thomas Harriot who had proposed the existence of atoms in his letters to Kepler who, for his part, refused to believe him. This did not stop him, however, from expounding the idea in his booklet, without as much as acknowledging the correspondence with Harriot.[8]

This is where Kepler's reflections on snow ended. Try as he might, our hero found no satisfactory explanation for the hexagonal shape of snowflakes. After pages and pages of deliberations he raised his hands in defeat and posited the existence of a *facultas formatrix,* a "formative faculty," that assembles snowflakes in their beautiful pattern according to the Creator's design. Deep down he knew that this was a cop-out, and in the last paragraph of the booklet he challenged future scientists to find the underlying reasons for the snowflakes' hexagonal patterns. In particular he "knock[ed] on the door of chemistry" and predicted that chemists would eventually be able to give the answer. This was to happen only three centuries later.

So Kepler was unable to give satisfactory answers to the questions he had set himself. But his booklet is remarkable nevertheless. In the course of his investigations he made extraordinary comments about close packings in two and three dimensions. He hinted at the fact that the hexagonal packing of two-dimensional spheres is the densest packing possible. He offered no proof, and it was to take 341 years to confirm this statement conclusively.

---

[8] Harriot had also had all the insights into crystalline order that have been attributed to Kepler. He had also recognized the similarities between HCP and FCC packing.

Then he stated as a truism that the arrangement described previously is the densest packing in three dimensions. We know that it took a total of 387 years to prove that conjecture.[9]

Let us return to snow. One must differentiate between snowflakes and snow crystals. Kepler entitled his treatise *Strena seu de nive sexangula,* which translates as "New Year's gift, or [a treatise] on six-cornered snow." In 1966, Oxford University Press published the first English version of this work and mistranslated *nive* as *snowflake.* However, snowflakes are hodge-podge aggregates of several snow *crystals,* and only the latter are six-cornered.[10] Clearly Kepler was talking about them.

So why are snow crystals six-cornered? Kepler hinted at two possible reasons. One, the hexagonal arrangement of circles is the densest possible arrangement of spheres in the plane. Two, hexagons can tile a plane without leaving gaps. His first explanation is completely off the mark. It would have been ever so cute, but the snow crystal's shape has absolutely nothing to do with dense packings in two dimensions.[11] His second explanation does have some basis in reality, as we shall see.

A quarter of a century after the publication of Kepler's New Year's gift for his friend Wacker von Wackerfels, the French mathematician and philosopher René Descartes (1596–1650) also decided to take a close look at snow crystals. He provided quite accurate descriptions of these "little plates of ice . . . so perfectly formed in hexagons, and of which the six sides are so straight, and the six angles so equal, that it is impossible for man to make anything so exact." Another twenty years went by until the experimentalist Robert Hooke (1635–1703) did him one better. Hooke did not just use the unaided eye to observe nature in the small, he spent his career peering at anything that would sit still under the most modern, state-of-

---

[9] Apart from the unproven hypotheses, the treatise also contains a blooper. At one point Kepler claims that if a space is completely filled with cubes of equal size, then *"unum cubum contingunt alii . . . octo et triginta"* (one cube is touched by thirty-eight others). Thirty-eight is way, way off the mark. Readers may recall the once popular "Hungarian Cube" that was invented by the design professor Rubik. It is made up of three layers of cubes, each layer consisting of three rows, with three cubes in each row. This gives a total of $3 \times 3 \times 3 = 27$ cubes, that is, 1 in the middle, and 26 arranged around it. Looked at in another manner, a cube in the center can be touched at its faces by 6, at its edges by 12, and at the corners by 8 other cubes. Again, this results in a total of $6 + 12 + 8 = 26$ cubes touching a central cube, and not 38, as Kepler, the Imperial Mathematician, claimed.

[10] However, only the title of the book is misleading. The chapter explaining six-corneredness is entitled "On the shape of snow crystals."

[11] More will be said about close packings in the plane in the next chapter.

the-art gadget of the day, the microscope.[12] Of course, snow crystals did not escape his attention, and in his bestseller, the *Micrographia* (1665), he included many drawings of these complex and intricate marvels of nature.

Nothing much happened with respect to snow crystals until the 1920s, when the American farmer and photo enthusiast, Wilson A. Bentley (1865–1931), bettered Robert Hooke's *Micrographia*. The Virginia ranchowner became a specialist in micro*photog*raphy. Holding his breath so the subjects would not melt, he captured some five thousand snow crystals on film. With his pictures he proved the saying—albeit not in the mathematical sense—that no two snowflakes are ever alike. "Every crystal was a masterpiece of design and no one design was ever repeated," he wrote. "When a snowflake melted, that design was forever lost. Just that much beauty was gone, without leaving any record behind"—except if it had had its picture taken before its demise by William "Snowflake" Bentley.

At about the same time, beyond the Pacific Ocean, the Japanese nuclear physicist Ukichiro Nakaya (1900–1962) undertook the first systematic study of snow crystals. He had been offered a chair at Hokkaido Imperial University where he felt overqualified and underemployed. There was simply no nuclear research going on in Hokkaido. So he had to seek another field of interest and turned his attention to something lower tech: snow. Nakaya began to grow artificial snow crystals under controlled laboratory conditions and could soon confirm that they are never the same shape. He was able to classify the forms, however, and developed the *Nakaya diagram,* which relates the shapes of snow crystals to meteorological conditions.

After snow crystals were observed by Kepler and Descartes with the naked eye, by Hooke with the microscope, by Bentley with a camera, and by Nakaya in a cold lab, it became time to use more advanced machinery. Back in Germany, the physicist Max von Laue had discovered the diffraction of X-rays by crystals. This achievement, which earned him the Nobel Prize in 1914, allowed the inspection of a snow crystal's structure and shape at the molecular level. The results were worked out in the late 1920s and they finally provided answers to the questions that Kepler had posed three centuries earlier.

Ice is frozen water and, as the word indicates, snow crystals are *crystallized* water. Water, in turn, is composed of H_2O molecules that contain two hydrogen atoms and one oxygen atom. But in the early seventeenth century nobody knew about the teeny-weeny building blocks that make up all matter—of which four million can be placed next to each other along the

---

[12] Nature in the large was observed with telescopes since the times of Harriot and Galileo.

length of 1 millimeter. Thomas Harriot did think of atoms as a convenient way to explain weight differences between various materials and even wrote about it to Kepler. He put forward the thesis that the atoms of light materials are packed in such a manner that the central one is touched by six others, while the atoms of heavy materials are surrounded by twelve others.[13] Little did Harriot know that this is, in fact, a partial explanation for the mass of different materials.

But this model was just a construct that came in handy as an explanation. Kepler himself refused to entertain the idea that all matter—stones, plants, and even human beings—could be made up of these little building blocks.[14] In 1805 the English chemist and physicist John Dalton of Manchester once again took up the hypothesis that matter is composed of small particles, in order to explain certain phenomena that occur during chemical reactions. But for a long time atoms remained no more than a hypothesis. Only by the end of the nineteenth century was it generally accepted that matter is composed of atoms that combine to form molecules. With the advent of electron microscopes—with a resolution of nearly one millionth of a millimeter—toward the middle of the twentieth century, molecules could finally be seen.

Kepler thought of snow crystals as aggregates of globules of condensed moisture. Again he was wrong. Snow crystals are composed of water that crystallizes around dust particles that have been carried up into the atmosphere. They are not always six-cornered, though, and their exact shapes depend on the temperature. Apparently the snow crystals that Kepler observed had formed when the air temperature in the clouds was between −12 and −16 degrees centigrade because this is when they take on hexagonal shape.[15] As the crystals grow they become heavier and fall toward earth. On the way down, up to two hundred of them may become bunched together to form snowflakes.

The shape and form of a crystal reflects the way in which its molecules are arranged. And the arrangement of the molecules, in turn, depends on how the constituent atoms are fused together. Snow crystals are six-cornered because of the manner in which the water molecules are arranged. H_2O molecules roughly resemble tetrahedra, with the oxygen atom sitting in the

---

[13] Apparently Harriot anticipated Kepler's conjecture by a few years, and it could have been called *Harriot's conjecture* with at least equal justification.

[14] As we saw, this did not stop him from being the first scientist ever to formulate the proposition that a macroscopic feature of an object may have roots so tiny that they cannot be seen by the unaided eye.

[15] When it is colder, ice crystals form columns or they are star shaped. If the temperature is slightly warmer, plates form, then columns, needles, and finally plates again.

middle, the two hydrogen atoms perched at the end of two vertices, and the other two vertices being occupied by unbounded electron pairs. Molecules can link themselves together, with the hydrogen atom of one molecule attaching itself to an unbounded electron of another water molecule. This is called a hydrogen bond.

Different shapes can be built using hydrogen bonds, so why should water molecules choose a hexagonal arrangement when forming snow crystals? The answer lies in a universal principle of nature that says that *any physical system, left to itself, strives to achieve a state in which energy is at its lowest level.* This principle is accepted as an axiom, which is a fancy word for something that is obviously true but nobody really knows why. Water, snow, and ice are no exceptions to the principle of lowest energy, and the arrangement of the molecules will be selected accordingly.

Since heat is a manifestation of energy, different shapes will appear at different temperatures.[16] At relatively high temperatures water molecules are too busy swirling around to get attached to each other. But with falling temperatures, they calm down and start to bond. At temperatures between −12 and −16 degrees centigrade, the lowest energy level is achieved when the H_2O molecules are organized in a lattice such that each molecule is surrounded by four neighbors: one molecule sits at the center, and the other four at the corners of a tetrahedron that surround the central molecule.[17] Viewed from the top, this arrangement appears as a hexagon. And this is whence six-corneredness arises.

When ice crystals start to grow around the nucleus, they initially retain the hexagonal shape of their molecular structure. As more water molecules travel through the vapor-filled air, looking for a good place to land, the swirling hexagons offer themselves as ideal airports. And since the corners of the hexagons stick out farther into space than the edges, this is where the molecules like to dock. With more and more molecules attaching themselves to each other, tree-like structures grow out of the corners of the hexagon. This is why snow crystals are flat and six-cornered.

We see that six-corneredness has nothing to do with the fact that placing circles in a hexagonal arrangement represents the densest possible arrangement in a plane. In fact, the packing of atoms in ice crystals is not particularly dense, as can be verified by the following simple experiment. Fill a bottle to the brim with water and then put it in your freezer. The bot-

---

[16] Even freezing temperatures are considered heat, as long as they are above the absolute zero of −273 degrees centigrade, or zero degrees Kelvin.

[17] This tetrahedron derives from, but is not identical to, the tetrahedron that is formed by the H_2O atom itself.

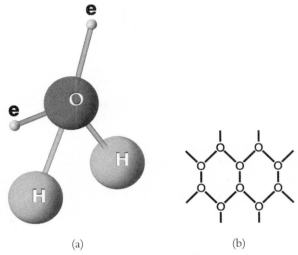

(a)   (b)

An H_2O molecule (a), and a crystal lattice, hexagonal arrangement
viewed from the top (b)

tle will crack. Why does this happen? After all, there are as many H_2O molecules in the bottle before and after freezing. The bottle breaks because the molecules use up more space when they are arranged as a lattice of ice than when they are in liquid form. In other words, the hexagonal arrangement of molecules in ice is not as dense as the arrangement of the molecules in water, so it certainly can't be the densest packing possible.

On the other hand, Kepler's suspicion that a relationship exists between the six-corneredness of snow crystals and the tiling of the plane turns out to be correct. Regular crystals are made up of so-called unit cells of a specific shape, which are repeated over and over again. Imagine a wallpaper pattern. If you look closely you will be able to pick out the motif which is representative of the pattern, and which—when repeated upward and downward, right and left—generates the wallpaper pattern. Now if the motifs are limited to regular polygons—geometric figures whose sides and angles are equal—then only certain shapes are possible. In fact, of all imaginary shapes there are only three regular polygons that can be used as wallpaper motifs. The motif could be a square because you can fill wallpaper or cover a bathroom floor with square tiles without leaving any gaps. The motif could be a triangle, for the same reason. Or it could be a hexagon. But that's it. The motif can never be a pentagon, for example, because try as you might, you will not be able to arrange a floor with pentagonal tiles. Gaps appear between the pentagons that cannot be filled.

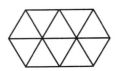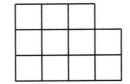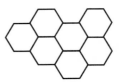

Wallpaper: triangles, squares, and hexagons

In this sense crystals are like wallpaper. When the lattice of a crystal is viewed from above, its shape must be extendable in all directions without leaving any gaps. The wallpaper argument narrows down the possibilities significantly. In the 1780s Robert Bergman peered at crystals using the recently invented microscope. His investigations suggested to him that they consist of packed rhombohedra. During the first half of the nineteenth century, the German mineralogist Johann Hessel (1796–1872) and the French astronomer and physicist Auguste Bravais (1811–1863) provided mathematical proof that only thirty-two classes of crystal forms are possible. They also showed that if a crystal was to possess n-fold symmetry, then n could only be 3, 4, or 6.[18] That means that a crystal lattice can be composed only of triangles, squares, or hexagons. No other regular shapes fit the requirements.

Kepler knew, of course, that a kitchen floor can be tiled with one of three different shapes: triangles, squares, or hexagons. But while he thought he knew why bees choose hexagons over the other two shapes, he was never able to figure out why snow crystals do. Exasperated, he left the matter at that. As we now know, the way out of his dilemma—trilemma actually—is provided by the principle of lowest energy. It is because the energy level of snow (at temperatures between −12 and −16 degrees centigrade) is lowest when H_2O molecules are arranged in a hexagonal arrangement that snow crystals have a six-cornered shape.[19]

---

[18] Twofold symmetry, which corresponds to symmetry in the plane, is also possible, in principle, and *quasi*crystals have fivefold symmetry.

[19] Christoph Lüthy of the Center of Medieval and Renaissance Natural Philosophy of the University of Nijmegen has advised me of the following: there exists some evidence that Kepler may have also been inspired by his reflections on the closest packing of spheres by Giordano Bruno's 1588 publication *Articuli Adversus Mathematicos*.

CHAPTER 3

Fire Hydrants and Soccer Players

In this and the following chapter we limit the discussion to two dimensions, that is, to the plane. And in the plane—imagine a tabletop—we want to arrange two-dimensional spheres, or circles, in an efficient manner. The obvious question is, How can disks of the same size be arranged so that the density in the plane is maximized? In chapter 1 we computed that six circles, arranged in a hexagon around a central circle, achieve a density of 90.7 percent. One can easily verify that this is the closest possible packing in two dimensions, just by pushing the coins around. But does this *prove* anything? Obviously not! In this and the next chapter we shall show that the hexagonal arrangement is, in fact, the densest packing possible.

But before we set out to analyze this question in more detail let us mention another problem, the so-called *dual*. What is the smallest number of circles (all of identical size, of course) that is needed to completely cover a plane? In contrast to the by-now-familiar "packing problem" this question is called the "covering problem."[1] In the packing problem circles are allowed to have gaps between them but there must be no overlaps; in the covering problem circles are allowed to overlap, but there must be no gaps. Phrased a bit differently, in the packing problem the *gaps* must be minimized, in the covering problem the *overlaps* must be minimized. Phrased differently still, the density of a packing is *at most* 100 percent: no part of the plane may be covered by more than one disk. In the covering problem, on the other hand, the disks cover the whole plane, and then some. The density is *at least* 100 percent.

Imagine a group of dictators—this example comes from a very important mathematics treatise—whose powers reach the same distance in all

---

[1] A dual problem is, in some sense, the opposite of the original problem. The dual to minimization is maximization, the dual to the packing problem is the covering problem.

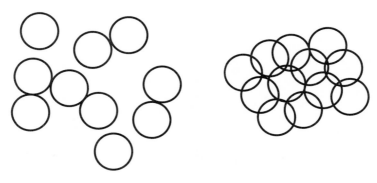

Packing (left), covering (right)

directions. As a consequence they each reign over a circular region. Since they do not wish to get into each other's hair they decide at one of the annual dictators' conferences to position themselves as far away from each other as possible. So the question they put to the Dictatorial Mathematician is, How can the largest number of dictators be packed into a certain area? By the way, any solution to this problem allows some fortunate individuals to remain in the pockets of democracy between the circular dictatorships.

The dual to this packing problem is, for example, how municipal services have to be distributed so that everyone in town gets to enjoy a certain minimal level of service. How many garbage containers must be placed so that no resident needs to carry the trash more than 50 meters from the front door to the closest bin? How many fire engines are needed, and where must they be located, so that nobody must wait more than six minutes after a distress call? Note that here, too, there are some lucky individuals who have more than one garbage bin to choose from. If the citizen is really lucky, fire engines from two stations will appear at the front door within six minutes after Junior sets the house on fire.

Let us move from the sublime to the ridiculous and consider carpet bombing. "Carpet" is, of course, a misnomer, since—apart from the deceptive evocation of domestic coziness—it gives the impression that the area to which a bomb inflicts damage is rectangular. But shrapnel flies the same distance in all directions, hence the ravaged region is circular. It is a consequence of modern warfare that a number of good explosives partially go to waste if "scorched earth" is to be achieved because of the overlaps.

On a more banal note, packing problems in two dimensions arise when containers are loaded onto the deck of a ship, when tailors cut shapes from a piece of cloth, or when barrels are stacked along the walls of a wine cellar. Covering problems arise in command and control systems in the military, monitoring of airspace, distribution of a sales force, or allocation of service technicians. The list of examples could be lengthened, but then again no mathematician worth his salt worries about applications. Hypotheses are put forth, solutions are discovered, and theorems are proved purely for their own sake, rarely for an ulterior purpose. Sometimes applications may appear later, sometimes much later, but this is not what makes the minds of mathematicians tick. For example, the work of nineteenth-century mathematician Arthur Cayley on algebraic matrices served as a foundation for quantum mechanics, which was developed only in the twentieth century. Cayley did not know what his matrices would be good for, and he didn't care.

Now, the best imaginable packing would achieve a perfect density of 100 percent. The most economical covering would also achieve a perfect density of just 100 percent. Can perfect packings or coverings ever be achieved? Not with circular disks, since they either leave gaps or they overlap. But why should we limit ourselves to disks? And even if we do limit ourselves to regular shapes, there may be other geometrical objects that can do the trick. Not the pentagons—one hundred percent covering of the plane by pentagons is impossible. Try as one might, there is no way to place them next to each other without leaving gaps in between. So do packings or coverings of 100 percent exist?

The answer is yes. But there is a surprise. Not only are there shapes that permit perfect coverings and shapes that permit perfect packings, they are the exact same shapes. Perfect covering and perfect packing go hand in hand. An arrangement that achieves a packing with a density of 100 percent achieves a perfect covering at the same time. Can you imagine an arrangement of geometrical objects that is, at the same time, both the densest packing and the most economical covering? Let us take a step back. In chapter 2 we saw that Johannes Kepler had already established that triangles, squares, and hexagons are able to *tile* the kitchen floor. Hence these shapes both pack and cover the plane perfectly. And this is exactly the definition of a tiling: an arrangement of shapes that both covers and packs perfectly. The three shapes mentioned—triangles, squares, and hexagons—are the only regular shapes that can do that. That's why ceramics shops usually do not carry pentagonal, heptagonal, or other exotically shaped tiles in their inventories.

We begin by stipulating that the centers of the circles must lie on a regular grid (also called a lattice). Let us look at a city like Manhattan and consider the fire department's problem of where to place fire hydrants. It is of paramount importance that total coverage be achieved, that hoses attached to the hydrants can reach every nook and cranny of the city. Hence the planners are faced with a covering problem. Since the lengths of the hoses determine how far they can reach in all directions, the hydrant's coverage is anything inside the circle of a radius equal to the hose's length. So we have a circle covering problem.[2] But there is more. If a blaze erupts we do not want fire fighters running around frantically with their hoses, looking for water outlets to attach them to. It would be a good idea if the hydrants were placed in such a manner that they are easily found, even when chaos and confusion reigns. So the fire chief may decide to place hydrants only at intersections of certain blocks and avenues. The possible locations of the hydrants form a grid. What we have, therefore, is a circle covering problem in a two-dimensional lattice. The solution could be, say, to place hydrants at all intersections of all avenues and even-numbered streets.

A nonlattice covering, on the other hand, can be portrayed if we look at soccer players on a playing field. Each captain wants his team to defend its side of the field in its entirety. Every player is responsible for his part of the field, say, 20 meters in all directions. Hence he must defend a circle with a 20-meter radius. But while we do expect the goalkeeper to be posted somewhere near the goal, the other ten players do not have to stand at regular intervals along the field. This leaves them more options and the players of the opposing team have greater difficulty keeping track of them. Hence the soccer players are faced with an example of a nonlattice circle covering problem. The reader is encouraged to conjure up further examples of packing and covering problems, both of the lattice and nonlattice variety.

In the first problem (the lattice variety), restrictions are put on the possible solutions. So when a contention has been proven for a lattice, the general problem has not yet been solved by any means. It is quite conceivable that a nonlattice solution is lurking around somewhere that is superior to the lattice solution. On the other hand, it could turn out that the solution to the lattice problem is also the optimal solution to the general problem. But this requires a proof. It may be—and often is—much harder to find the solution to the nonlattice variety of the problem, since it allows more possibilities.

Who was the first person to put his thoughts about the densest arrangement of circles to print? Generally it is assumed that it was Kepler in the treatise *The Six-Cornered Snowflake* (see chapter 2). But while researching this book I found otherwise. In Zürich I was invited to the office of Caspar Schwabe, a designer of geometrical objects. Sitting among the strange mathematical objects that Schwabe had either produced himself or col-

---

[2] If we consider that the hoses must reach a building's upper floors, we are faced with a three-dimensional covering problem. But in this chapter we ignore the third dimension.

lected—among them an original *Tensegrity* by Buckminster Fuller that takes up most of the airspace in his office—we made a small discovery. Schwabe showed me a recent acquisition of his, a richly illustrated book by the Renaissance artist Albrecht Dürer. I here use the word "book" with a certain nonchalance, since printing and typography in Dürer's day was a very recent invention. Dürer was born in 1471, while printing technology, invented by the metallurgist and goldsmith Johann Gutenberg just thirty years earlier, was still in the alpha testing stage. Throughout his professional life, Dürer would put this new technology to good use.

Albrecht was number three in a line of children that ran to the number eighteen. His father was a jeweler who had immigrated from Hungary to the German city of Nüremberg. He was a religious man who managed to instill a healthy fear of God in his children. Dürer described him as "gentle and patient . . . friendly towards all and full of gratitude to his Maker."

Albrecht Dürer (self portrait)

Albrecht received his initial training in painting and woodcutting in Germany. After finishing his apprenticeship, which he spent wandering through German cities, his parents arranged his marriage to a young woman of good family, Agnes Frey. Albrecht himself was not very keen about getting married and was not too happy in matrimony. A few months after his wedding he took off again, this time to Italy. In Venice he hung out with local artists, visited galleries, and sketched scenes of the city. Upon his return to Germany in 1512 he found himself in the employ of Kaiser Maximilian.

Unfortunately, over time celebrity went to Dürer's head and he became something of a prima donna. As a consequence he also became a bit neurotic, and on his next trip to Italy he thought that aspiring competitors were trying to get rid of him by putting poison in his food. He was so scared of this perceived threat that he would not accept any invitations to dinner. Back in his beloved Nüremberg he and his wife were among the first to make a living off the new technology of typography, which had by now reached the beta testing stage. The Dürers set up their own printing press. Albrecht supplied the input and Agnes sold the output at the local fairs. Today Dürer is considered one of the foremost painters and woodcutters of all times.

Dürer's mathematical interests and achievements are less well known, but as a true Renaissance man he was well-versed not only in the arts but also in the sciences. At one time or another he had studied such diverse fields as anatomy, aeronautics, and architecture. With some pride he considered himself a *Mathematicus,* and gave himself that title in the preface of the book I looked at in Zürich. Dürer considered mathematics a necessary prerequisite for all the arts. The problem of perspective, for example, had always vexed painters, and up until the fifteenth century they depicted figures only in a single plane. In the 1420s the Florentine architect Filippo Brunelleschi invented the technique of perspective drawing. Dürer, who had learned about it during a trip to Italy, was one the first artists to depict figures simultaneously in the fore- and background.

His geometrical investigations were pathbreaking, and his book *Unterweisung der Messung mit Zirkel und Richtscheit* (Instructions on measuring with compass and ruler) was the first mathematics textbook in German. In it he discussed perspective and proportions and showed how to construct figures with ruler and compass.

Dürer's geometrical investigations influenced not only artists but also mathematicians. His attempts to overcome the problems of projection, perspective, and depiction of moving bodies led him to the development of a new discipline, *descriptive geometry*. This branch of geometry was put on a sound mathematical basis in the second half of the eighteenth century by the Frenchman Gaspard Monge. Recently the investigations begun by Albrecht Dürer saw a renaissance when computer scientists started dealing with the representation of three-dimensional objects on two-dimensional screens. They had to deal with the problems of hidden lines and—once again—of perspective.

The book we looked at was published in 1528, shortly after the artist's death. Leafing through the 450-year-old tome, we stumbled upon a pair of pictures that immediately caught our attention. In the first one, nine circles are packed in a quadratic arrangement, and next to it is a figure of seven full circles and ten partial circles packed hexagonally in a square of similar size. The picture corresponds nearly exactly to a figure in a modern introduction to geometry.

It took a while to decipher the ancient German text that accompanied the figures, but it finally emerged that Dürer did indeed refer to packings, although he made no reference to densest arrangements. What he wrote was that there are only two ways to decorate the ceiling or the walls of a house with a regular arrangement of circles: the square packing and the hexagonal packing. He then implies that the latter is denser than the former. It cannot be excluded, and we may even assume, that the well-read Kepler actually perused Dürer's book in the late 1500s or early 1600s and let himself be inspired by it.

So Dürer pointed out that among the two regular arrangements for circles the hexagonal one is *denser* than the quadratic, and Kepler claimed that the hexagonal packing is the *densest* arrangement possible in two dimensions. But who proved this? As was mentioned previously, this question must be separated into two parts. It took two and a half centuries from the date of Dürer's publication to prove that the hexagonal arrangement is the densest lattice packing of circles in the two-dimensional plane. It then took another 170 years to show that this same hexagonal arrangement is also the densest general packing. In the remainder of this chapter we investigate the densest lattice arrangement. The densest general arrangement will be discussed in chapter 4.

The action shifts to Italy. Giuseppe Lodovico Lagrangia (who later called himself Joseph-Louis Lagrange) was born in Turin in 1736 to a civil servant of good social position who was in charge of the treasury of the city's Office of Public Works and Fortifications. Maybe this worthy man had played around a bit with the content of the treasury in order to fortify himself rather than the city, or maybe he was just a shrewd investor. Whichever, he became a wealthy man. But unfortunately goddess Fortuna's favors did not last forever, and one day Lagrange senior's luck ran out: he lost everything in speculations. Young Giuseppe Lodovico was shipped off to college to study law. As a seventeen-year-old student he excelled in classical Latin (as did Kepler 160 years before him), though he showed little interest in mathematics. But one day he happened upon a mathematical treatise, started reading it, got excited, and decided to devote the rest of his life to the subject. He immediately immersed himself day and night in intense study of mathematics. It was fortunate for the world that his father's financial luck had run out, because—as Lagrange himself said later in life—had he been rich, he would have become a lawyer.

Two years later Lagrange was ready to establish his early fame. He wrote a letter to the then grandmaster of mathematics, Leonhard Euler, from Switzerland. Euler, director of mathematics at the Academy of Sciences in Berlin, was one of the most prolific mathematicians who ever lived and made contributions to all areas of mathematics.

In his letter Lagrange described the solution to a problem that had vexed mathematicians for more than half a century. It was the so-called isoperimetric problem, which consisted of maximizing a surface, or a volume, in the presence of constraints and boundary conditions. We already met an example of the isoperimetric problem in Kepler's treatise on six-cornered snow. Remember the bee whose task it is to maximize the storage space in the honeycomb while minimizing the amount of wax that is needed to build it? That's an example of an isoperimetric problem, and as Pappus of Alexandria had already claimed in the third century, the solution is the hexagon.

Actually, the first known example of an isoperimetric problem stems from much earlier, namely from the eighth century B.C. It is the story of Dido, queen of Carthage. According to legend, Dido fled from her tyrannical brother, King Pygmalion. She landed on the coast of North Africa and declared her intention to purchase a parcel of land. The locals were quite willing to take her money but weren't so keen on giving her much real estate. A snickering native told Dido she could take as much territory as she could cover with the hide of a bull. Apparently he thought of selling the beautiful lady a little garden of about 5 square meters, but the clever Dido had the last laugh. She cut the bull's hide into very thin strips, attached them end to end, and then announced that all territory that she could surround with this strip would be hers. She probably confused "covering" with "encircling" on purpose, but the poor locals fell for it anyway. After Dido had hit on the idea of cutting the bull's hide into strips, she still had to decide what shape the territory should have. It is here that the isoperimetric problem makes its appearance. Obviously Dido wanted her new home to extend over the greatest surface. Hence she had to figure out which shape would cover the greatest area, while being surrounded by a border of a given length. The answer is—you guessed it—a circle. As the locals watched in horror, Dido proceeded to encircle a large parcel of land with the strip and became queen of the round city of Carthage.

Lagrange wrote to Leonhard Euler about the three-dimensional version of Dido's problem, that is, the problem of finding the greatest volume to a fixed surface area. Euler had just been about to solve the problem himself, but realized that the unknown youth from Turin had found a method, later to become known as the *calculus of variations,* that was superior to the one he was going to propose. Recognizing the young man's gifts, he held back his own manuscript in order to get Lagrange's work published, and in one fell swoop the hitherto unknown Italian student established himself at the cutting edge of the profession. A little later, while investigating vibrating strings and the propagation of sound, Lagrange pointed out both an error that the great Isaac Newton had made when he analyzed the same phenomenon and a lack of generality in the methods used by previous scientists. He went on to write a pathbreaking discussion of echoes, beats, and compound sounds, thereby establishing himself as an early fan of rock music, two hundred years before the Rolling Stones hit the charts.

But this was not the Woodstock generation, and rock music did not automatically mean pacifism. So at age nineteen Giuseppe Lodovico was appointed professor of mathematics at the Royal Artillery School of Turin. After all, not only artists needed a solid grounding in mathematics, as Dürer had pointed out; shooters too had to be able to compute where their shells would fall. While teaching artillery officers to do their math, Lagrange continued his research, and at the age of twenty-five he was already recognized as the foremost mathematician of his time. But the intense work of the previous years took its toll and Lagrange fell ill, physically and psychologically. He eventually recovered from his bodily ailment but his nervous system never fully recuperated. Throughout his life he would suffer from bouts of deep melancholy and depression.

Lagrange received many offers of work in various countries, but the shy young man turned them all down, preferring to live in modest circumstances in his native Turin. He wanted to devote himself entirely to mathematics. In 1766 the situation changed, however. Leonhard Euler was fed up with his boss, Frederick the Great, who insisted on meddling in the matters of the Academy. He left Berlin to set up camp in St. Petersburg with Peter the Great. (No relation, they just both considered themselves great.) Frederick the Great was greatly angered by the desertion and suggested to

Lagrange that "the greatest mathematician of Europe" join up with "the greatest king of Europe." Even the unassuming Lagrange could not refuse such a proposition and, on November 6, 1766, he became Euler's successor as Director of Mathematics at the Berlin Academy of Sciences. He would spend the next twenty years at Frederick's court.

During his Berlin years, Lagrange produced about one memoir a month, altogether hundreds of papers. He was no narrowminded crank, mind you, but dealt with a varied array of subjects: algebra, the theory of numbers (he proved some of Fermat's theorems, but alas, not the celebrated one), analytical geometry, differential equations, astronomy, and mechanics. While at the Prussian court he also walked away with most of the prizes that the Académie des Sciences de Paris offered biannually for the solution of a particularly interesting problem. (He won in 1764, 1766, 1772, 1774, and 1778.)

Frederick died in 1787, and with this Lagrange's life in Berlin became less fun. He received numerous job offers, the most enticing one from Louis XVI, king of France. It contained, most important, a clause that freed him from all teaching duties, and Lagrange willingly accepted. The Académie des Sciences de Paris immediately made him a member—probably to let other young mathematicians finally take a shot at winning the prizes.

The beginning of the French Revolution saw Lagrange busy as a member of the Committee to Standardize Weights and Measures. The Académie, along with many other learned societies, had been forced to close its doors during the Reign of Terror, but the weights and measures commission was allowed to continue its work. Even the revolutionaries realized that some order was needed in the chaos that reigned in the markets. Merchants in different parts of the country used different modes of measurement (albeit keeping the price constant, which is a sure recipe for inflation), and the committee's brief was to devise a system that would be binding throughout the country.

After many hours of deliberation they came to the conclusion that the simple folk would find it easier to count and multiply if they could use their fingers. So they devised the decimal system and based the meter, the gram, and the liter on it. In contrast, the Romans had divided the pound and the foot into twelve ounces and twelve inches, presumably because they used all their fingers plus two toes to count up to a dozen. Until recently, even the English and Americans insisted that using fingers and a couple of toes to do their arithmetic was a good idea. It appears that the Anglo-Saxon folk were not so concerned with multiplication of weights and measures, but with division. And since a dozen can easily be divided by 2, 3, 4, and 6, while the metric system allows easy division only by 2 and 5, they adopted the dodecimal system. Only now are they slowly coming around to the system proposed by Lagrange and his men.[3]

The Reign of Terror did not leave Lagrange unscathed. A law was passed by the Assemblée Nationale that ordered the arrest of all foreigners born in enemy countries and the seizure of their property. This decree obviously applied to Lagrange—so much for the claim that he was a Frenchman—but the chemist Antoine-Laurent Lavoisier, one of the foremost French scientists of the time, and a colleague from the Weights and Measures Committee, intervened on the famous mathematician's behalf and an exception was made. So Lagrange was spared, but shortly thereafter Lavoisier—until then a respected government official and the Commissioner of Gunpowder—fell afoul of revolutionary ideology. A court took just a day to try, condemn, and execute this great man. Lagrange was deeply grieved by the killing of the friend who had saved him from arrest and possibly from a similar fate just a year earlier. "It took only a moment to cause his head to fall and a hundred years will not suffice to produce its like," he exclaimed.[4]

Privileges were severely frowned upon by the revolutionaries and without much ado Lagrange's "no teaching" clause was canceled. From then on he was forced to lecture, first at the École Polytechnique, today the top school for France's elite civil servants and business leaders, and later at the École Normale, which was founded as a school for teachers but is today the place for the real intellectuals. Not only did Lagrange have to teach, but the lectures were also taken down in shorthand so that the deputies could inspect for themselves whether the professor deviated from his subject matter. Apparently, the authorities were very worried about the subversiveness of differential equations and the counterrevolutionary influence of calculus and did not think of applying the first third of their battle cry (*"liberté, égalité, fraternité"*) to academic freedom.

In his later life, Lagrange was showered with honors for his achievements in mathematics. In 1796 the French commissary in Italy was dispatched to

[3] On July 4, 2000, Steven Thoburn, a greengrocer in Sunderland (England), committed a rather serious crime. He insisted on using pounds to weigh the bananas he sold, instead of kilograms, as prescribed by the laws of the European Union. Mr. Thoburn was convicted of breaking weights and measures legislation.

[4] A young bookkeeper who had been working for Lavoisier was so distressed by the death of his master that he picked up his belongings and left for America. He also took with him the know-how to make gunpowder. At the urging of President Thomas Jefferson the bookkeeper founded a company in Wilmington, Delaware, where his expertise on gunpowder production was put to good use. The young man's name was Irenée Du Pont. The rest is history.

the residence of his father to convey to him the republic's congratulations on the achievements of his son. After Napoleon came to power, Lagrange was named officer of the Légion d'Honneur and made a Count of the Empire. On April 3, 1813, one week before his death, he was awarded the Ordre Impérial de la Réunion. He is considered the foremost mathematician of the eighteenth century.

We now return to the subject of circle packings.[5] In 1773, while still at King Frederick's court, Lagrange wrote a treatise entitled "Recherche d'arithmétique," which was published in *Nouveaux mémoires de l'Académie royale des Sciences et Belles-Lettres de Berlin*. Lest the reader think that this treatise is a simple essay about the four arithmetic operations—addition, subtraction, multiplication, and division—be forewarned. Lagrange did not waste his time with trivial high school math. It was *higher* arithmetic—today called number theory—that he was dealing with. In the treatise Lagrange discussed *binary quadratic forms,* that is, expressions like $a^2 + 2b + c^2$. He determined that the most important characteristic of a quadratic form is its so-called *discriminant,* which is calculated as $a^2c^2 - b^2$. As it turns out, determinants have a close connection to grids (or lattices).

In order to explain the connection, we must first define the notion of a grid's *base.* Two vectors, one for each direction, determine a two-dimensional grid. In graphs and pictures these vectors are usually indicated by little arrows, and with their help travelers wandering through the grid can find their way around. The vectors and the angle between them form the grid's base.[6] In Manhattan, a useful base consists of the avenues and the streets—let's call it the "Manhattan base." The distance between two avenues (say 300 meters to the east) represents the vector in one direction and will be denoted by *a*. The distance between two streets (100 meters north) represents the vector in the other direction and will be denoted by *b*.[7] The angle between avenues and streets is 90°. Hence, an address like "8th and 12th" refers to the intersection of 8th

---

[5] The following pages are not easy reading. It is recommended that only mature readers, those who can withstand the graphic depiction of math, continue on. Readers more interested in the history of the sphere-packing problem may want to skip (or just skim) the next few pages and move directly to the conclusions at the end of this chapter. A different typeface is used here and elsewhere in the following chapters to indicate sections that deal more explicitly with math.

[6] For a three-dimensional lattice, the base consists of three vectors.

[7] The numbers and directions are just for illustrative purposes. The distances in New York City between streets and avenues are not exactly 100 and 300 meters, and the directions are not exactly north or east.

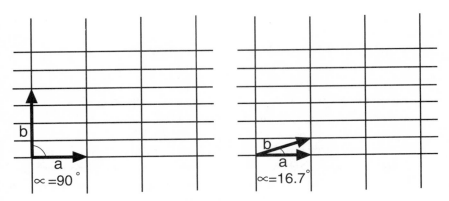

Two bases for Manhattan

Avenue and 12th Street. Starting from the imaginary point zero—somewhere near the Hudson River in our imaginary Manhattan—you would have to walk eight avenues ($8a = 2,400$ meters) to the east and twelve streets ($12b = 1,200$ meters) to the north in order to reach this location.

When Lagrange was doing his higher arithmetic there was no Eiffel Tower in Paris and tall buildings did not yet exist in Manhattan either. Thus there were several ways to reach the same spot. Starting again at point zero, one could start out in any direction, then take a shortcut through the fields, and again land at the aforementioned "8th and 12th." For example, Peter Stuyvesant could have first walked four avenues to the east, and then cut through the remaining blocks to get to the intersection. His base would be composed of the two vectors "300 meters in the east direction" and "316 meters toward the east-northeast direction."[8] The angle between the two vectors is about 17°. Using the "Stuyvesant base" a traveler can also reach every spot on the grid, and it therefore also forms a base for the Manhattan lattice. In the same manner one can think of infinitely many different bases for Manhattan. And this presents a problem, as we shall presently see.

What do quadratic forms have to do with lattices? The area contained between avenues and streets is known as a block. More generally, the block in any lattice is usually called the fundamental cell. Wouldn't it be nice if some simple expression for the cell's surface existed? Well, we're in luck. As we show in the appendix, the surface of a base's block is just the square root of

---

[8] The direction is not exactly east–northeast. By Pythagoras, the hypothenuse of a right-angled triangle with perpendicular sides 100 and 300 meters long is 316 meters.

the quadratic form's discriminant. That's the link between grids and quadratic forms.

All this gave the mathematicians who were working on packing problems an idea. To find the lattice that represents the densest packing, one would have to do the following: (1) Find the grids that allow four circles to be placed at the corners of the cell, without any of them overlapping and (2) from among those grids choose the one whose fundamental cell has the smallest surface.

Why would that solve the packing problem? First, one must realize that each cell of the lattice contains one full circle, provided the vectors are sufficiently long and the angle between them is sufficiently large. If you place four pizzas at the corners of a cell, the slices that are contained within the cell form a full pizza again—no matter how the cell is shaped. In general there will be two large slices for the grown-ups, and two small slices for the kids. Put together, the four slices will form a full Quattro Stagioni pizza.

So to achieve the densest packing, all we have to do is find the lattice that wastes the least amount of space between the pizza slices.[9] In other words, we must find the lattice whose fundamental cell has the smallest surface. But the surface of the fundamental cell is the square root of its discriminant. So things are looking bright: All we have to do is check out all the bases of all the lattices and chose the one with the smallest discriminant.

But we have a problem, remember? Not only are there many, many lattices, but for each lattice there exists an infinite number of bases. How could we hope to find the best packing if we have to check all lattices an infinite number of times? So the problem Lagrange considered was how to reduce the infinite number of bases that fit each lattice to one single representative. Basically, this procedure, which is called reduction, consists of finding, among all the bases that describe a specific lattice, the one with the shortest vectors.[10] While fiddling around with the parameters of reduced bases, Lagrange found something quite remarkable. He noted that the discriminant of a reduced base could never be smaller than $\frac{3}{4} a^4$. He further found that the angle between the two vectors of a reduced base must lie between 60° and 90°.

Now recall that the surface of the cell is equal to the square root of the discriminant. So what Lagrange had discovered was that the surface of the lattice's cell could never be smaller than $0.866a^2$. As long as the two vectors and the angle between them are large enough to fit circles at the gridpoints without overlap, one may play around with the parameters as much as one likes: The

---

[9] The cell must be large enough to accommodate the pizzas at the corners of the fundamental cell without overlapping. (Who would want to eat squashed pizzas anyway?)
[10] Lagrange showed that a reduced base must satisfy the inequalities $0 \leq 2b \leq a^2 \leq c^2$, where $b = ac \cos \alpha$ (α being the angle between the two vectors).

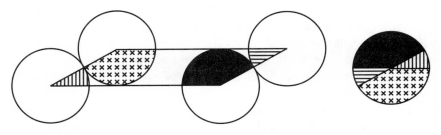

Pizzas in a cell

surface of the fundamental cell can never become smaller than that limit. And what kind of lattice attains this minimum? Lagrange showed that the minimum occurs only when $a = c$, and when the angle between the two vectors is 60°. And what, in turn, does that imply? Lo and behold, it's the hexagonal arrangement. Those square pizza boxes should certainly be on their way out: lots and lots of cardboard will be saved if the pizzas are placed into hexagonal boxes. (Round boxes would be better still, of course.) There is never anything new under the sun: Dürer and Kepler had already said so. But Lagrange did add the finishing touches.

❖ ❖ ❖

Lagrange provided the mathematical ingredients for the proof that the lattice whose points are arranged in a hexagonal manner minimizes wasted space. Actually, he had no interest in circle packings per se, and in his *Mémoire* of 1773 Lagrange considered quadratic forms purely as mathematical objects. He developed the theory of reduction of binary quadratic forms, but the connection to lattices escaped him. Only sixty years later, in 1831, did Carl Friedrich Gauss (of whom we will have much more to say

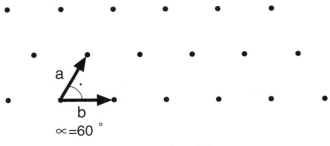

$a = c$, angle $= 60°$

in chapter 7) introduce lattices, and point out the connection between them and quadratic forms. But even he did not specifically consider circle packings. Toward the end of the nineteenth century, when interest in the subject resurfaced, it was realized that Lagrange had already provided all that was needed to show that the hexagonal packing is the densest arrangement of circles on a lattice.

In the next chapter we drop the requirement that the centers of the circles must lie on a lattice, and deal with the general problem of packings in two dimensions.

CHAPTER 4

Thue's Two Attempts
and Fejes-Tóth's Achievement

In the previous chapter we considered packings and coverings in two dimensions. We pointed out that there are actually two different questions that must be examined, depending on whether the centers of the circles lie on a lattice or not. Joseph-Louis Lagrange provided the ingredients for the proof that the hexagonal arrangement is the densest lattice packing in two dimensions. Now let us move on to the general case. Is there maybe an irregular arrangement that allows a denser packing?

It was long suspected that the hexagonal arrangement also represented the tightest general packing, that is, that there exists no nonlattice packing that is denser than the hexagonal arrangement. However, this is not quite obvious and must be proved. Take, for example, a two-dimensional region, say, a rectangle with a length of 6 and height 5.8. The rectangle's surface is 34.8. Using the square arrangement, six circles of radius 1.0 fit inside the rectangle, to give a density of 54.2 percent.[1] With the hexagonal lattice arrangement, on the other hand, seven circles can be arranged inside the rectangle, to give a density of 63.2 percent. That's better, but we can do better still if we don't require a lattice arrangement. By judiciously placing the circles we can fit *eight* of them inside the rectangle, to give a density of 72.2 percent. Does that mean that there are better arrangements than the hexagonal? The answer is: sometimes, if we consider limited areas. The reason for the higher density of the jumbled arrangement is that the rectangle does not extend to infinity. One must not confuse local density with global density; it is the latter we want to maximize. But this little example does point to a

---

[1] $6\pi r^2/34.8 = 54.2$ percent.

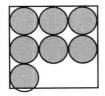 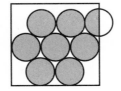 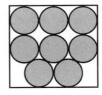

6, 7, and 8 full circles inside a square

problem, and the question must be asked whether there exist jumbled arrangements that are superior to the hexagonal lattice packing.

Toward the end of the nineteenth century, a Norwegian mathematician came along by the name of Axel Thue. Nowadays Thue is remembered mostly for his contributions to number theory, but he also did work in mathematical logic, geometry, and mechanics. His paper entitled "Om nogle geometrisk-taltheoretiske Theoremer" (About some geometric number-theoretic theorems) was the first published attempt to solve the general circle-packing problem. The stress is on the word *attempt,* because, as it turned out, it was no proof. In fact, it was not much more than an outline of a possible proof. Thue is sometimes awarded credit, nevertheless, for having been the first to solve the two-dimensional packing problem because, after all, who understands Norwegian?

Thue's scientific record is somewhat mixed. One well-known number theorist (Edmund Landau) judged a theorem by Thue to be "the most important discovery in elementary number theory that I know." But his biographers (Trygve Nagell, Atle Selberg, Sigmund Selberg, and Knut Thalberg) range in their assessment from "profound work, which started a new era in the theory of diophantine equations" down to "simple, but elegant and useful." Thue's life's work was generally fraught with frustrations; his assault on the circle packing problem was just one example.

Like Lagrange, young Axel came to the world of mathematics somewhat by accident. As a youth he was interested in physics and one day saw an advertisement for a book titled *Pendulum's Influence on Geometry.* What could a pendulum have to do with geometry, thought the boy, and ordered the book. After receiving it he realized that the advertisement had contained a typo: the real title was *Poncelet's Influence on Geometry.* Jean-Victor Poncelet was no pendulum. He was an engineer who took part in Napoleon's 1812 campaign and was taken captive by the Russians. During his imprisonment he wrote a treatise on analytic geometry, which wasn't published until fifty years later. So the book Thue received was pure mathematics, but he read on and became hooked. He also began a lifelong friendship with the author, Elling Holst, who would become his teacher and mentor.

Thue entered the University of Oslo and, as a student, gave more than a dozen lectures to select audiences at the Scientific Society and the Mathematical Seminar. Although obviously endowed with talent, the gaps in his education soon became apparent. Fellow Norwegian Sophus Lie, father of the appropriately named *Lie* algebra, remarked that Thue's "mathematical knowledge does not do justice to his gifts and his enthusiasm" and that "the probability of his succeeding in acquiring a sufficiently broad basis for his work diminishes year by year." But it was still not too late to fill the gaps, and Thue was granted scholarships for the years 1890 and 1891 to study in Leipzig and Berlin. He did not pass up the trip but took little advantage of the opportunity. Usually the brash young man was happy to just get the general idea of profound theories, postponing the study of the pertinent details for later. For example, he got off to a good start in Leipzig and was tutored by Sophus Lie. But when the time came to plunge into the nitty-gritty details, he went on a vacation trip to Prague instead. There he contracted yellow fever, which took him out of commission for a few months. The long and the short of it was that "my work . . . has not led to any positive or conclusive result," as he told a former teacher. In another letter he wrote, "I have written about 500 pages, but most of them can be scrapped." His association with Lie is the story of a missed opportunity. The rare chance to cooperate with the famous compatriot left absolutely no traces in Thue's work.

The near-total absence of citations to sources and of bibliographic material at the end of his papers—except for references to his own work—also indicated that Thue was simply not familiar with the pertinent literature. He should have realized that he was on the wrong path because, in his own words, "every time I came to a significant result, it turned out to be one well known already. It is to be hoped that there is still something left which can be said to be novel, and which learned mathematical palates will relish."

This haphazard approach continued throughout his career. Thue just did not enjoy immersing himself in the study of what other people had done before him. He followed a different approach: he reinvented the wheel over and over and over again. Only after he finished proving a theorem did he ask an assistant to check whether it was actually new, just to find out ever so often that someone had had the same idea before, sometimes dozens of years before. In a footnote at the end of his 1909 paper about the existence of transcendental numbers, Thue acknowledges that his proof was nearly identical to the one given by Liouville in 1851, more than half a century earlier. The title of another one of his papers, "Proof of a known theorem about

Axel Thue

transpositions," also attests to this lack of originality. Thue was fully aware of this foibles. "The reason I haven't published the work I have on hand is . . . because I didn't know what was new and what was already known." One possible reason for Thue's reluctance to study the professional literature was an inability to follow someone else's train of thought. Another explanation for this curious, self-imposed handicap could be that Thue was simply lazy. But he refused to learn from his mistakes and, instead, tried to make a virtue out of his character fault. He claimed that he didn't like to study a subject thoroughly because "such an investigation . . . has . . . an inhibiting effect on my imagination." As pretexts go, this must be one of the most disingenuous ones ever. Any schoolteacher has heard better excuses.

Before Thue decided to devote his life to reinventing wheels, he started his professional career with a seriously flawed paper. And this is the work that concerns us here. It was only Thue's second published paper, and we may ascribe the blooper to the twenty-nine-year-old's inexperience. Shortly after he returned from his trips abroad, at the 1892 annual meeting of the Scandinavian Society of Natural Scientists, Thue decided to give a presentation about his ideas on the general problem of circle packings.

He may have been aware of Lagrange's solution for lattices, though—true to form—he did not mention the Frenchman's work. But Thue's aim was to solve the general packing problem, without restricting the circles to a lattice.

At about the same time a young German by the name of Hermann Minkowski was beginning his rise to mathematical stardom. He was born in 1864 in Russia, and his family moved to the German town of Königsberg while Hermann was still a child. In school he became friendly with a slightly older boy, David Hilbert, who would later become one of the most influential mathematicians of the twentieth century (see chapter 8). But as a boy it was Minkowski who stole the show. In 1882, at age eighteen, he followed in Lagrange's footsteps by winning the Grand Prix des Sciences Mathématiques of the Académie des Sciences in Paris. The competition question concerned the representation of an integer as a sum of five squares, and the pupil from the Königsberg *Gymnasium,* the high school, was awarded the prize jointly with the English number-theorist Henry John Stephen Smith from Oxford.[2] Actually Minkowski nearly lost his prize again, because the English—fair sportsmen that they always are—didn't take kindly to the fact that the esteemed Savilian Professor of Geometry should have to share the honor with a mere boy from Germany. Smith himself was no longer in a position to care because he had died a few weeks earlier. But lawyers were put to work on the small print of the competition bylaws and they managed to find a reason to appeal against the Académie's decision: Minkowski's essay had been written in German, while the rules clearly stated that the paper was to be written in French. But the protests from the other side of the Channel were to no avail. The jury did not relent and Minkowski walked away with his half of the prize. It is ironic that nowadays the undistinguished Smith is mostly remembered for having shared the Grand Prix with the great Minkowski.

The young man's precocious career was interrupted when he had to serve in the army of his fatherland, but took off again after he became assistant professor at the Federal Institute of Technology (ETH) in Zürich. Minkowski, a rather shy man, was not a captivating speaker, and many students preferred to absent themselves from his lectures. One student, in particular, aroused Minkowski's ire because he constantly missed his classes. This student later went on to make a name for himself as a patent clerk in the Swiss capital

---

[2] Professor H. J. S. Smith used all three of his Christian names. Henry Smith would have sounded a bit too pedestrian for an Oxford don.

Hermann Minkowski

Berne. Apparently not all of the professor's teachings were lost on the truant student, however—his name, of course, was Albert Einstein—because Minkowski's invention of the "space-time continuum" laid the mathematical foundations for the theory of relativity. In 1902, at the instigation of his boyhood friend David Hilbert, who was by then a famous professor, Minkowski received a call to the University of Göttingen, then the world-center of mathematics. The two men became close collaborators until Minkowski's sudden death, at age forty-four, from a ruptured appendix.

Axel Thue's concern in 1892 was a line of research that Minkowski had invented. It would later be called the "geometry of numbers." Maybe it was Minkowski's Jewish background that kindled his interest in this new theory, which combined geometry and number theory. *Gematria* (a Hebrew-Yiddish term for "geometry") is a somewhat tongue-in-cheek technique in which each letter of the Hebrew alphabet is assigned a number. This then allows rabbis to interpret holy texts by their numerical equivalents.[3] But Minkowski didn't concern himself with holy texts; his line of inquiry was directed toward quadratic forms (which we met in the last chapter), and his version of gematria eventually developed into an independent branch of number theory. With its help, problems in number theory are studied by the use of geometric methods, or vice versa. Sometimes the solution to a problem that seems intractable in its original setting may become obvious in the context of figures. Minkowski's suggestion was to translate number-theoretic problems into geometrical terms, wherein proofs may be found more easily. Then they are translated back into number theory and, voilà, we have the solution.

As an example let us take an airline that wants to renew its fleet. How many jets should the company buy so that (1) its profits are maximized and (2) certain conditions are fulfilled: at least one landing a day in Paris, Rome, and London; no more than twelve weekly departures from Seattle; at least twice as many available seats on the leg between Atlanta and San Francisco as on the leg between Cairo and Tel Aviv; no takeoffs during the night in Zürich; and so forth. This is a problem of optimization under constraints, and computer programs have been developed to find the best possible number of aircraft.[4] After much cranking and churning, the computer may spit out the following result: profits are maximized with 17.9 airplanes. Obviously, this solution is no good. Neither Boeing nor Airbus nor any other company has devised a method to build and fly fractional planes. And don't you dare just round the number up or down. The closeness of the result to the number 18 should not fool you, because it is by no means certain that 18 aircraft is the optimal integer solution. Even if the best fractional result is 17.9, the optimum fleet could contain 6 aircraft, or 74, or 38!

So how can one know whether an integer solution to the problem exists? Let's say there are *d* conditions of the form "no more than twelve departures," "at least twice as many seats," and so on. These conditions define a system of *d* inequalities. Translated into geometrical terms, the system defines a body suspended in *d*-dimensional space. Now take a look at the axes of this space. In each direction they define a continuum of real numbers. The points on the axis that represent integer numbers define a

---

[3] The infamous "Bible Codes" provide examples for the uses and abuses of gematria.

[4] More on linear programming in chapter 12.

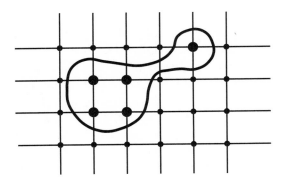

Body suspended in a lattice

lattice, or grid, in d-dimensional space. So the question becomes: Does a certain body that is suspended in a d-dimensional lattice contain at least one of the lattice points, or does it float between them? Translated back into arithmetic, this is equivalent to the question: Does a system of d inequalities (which defines the body) have an integer solution (contain a lattice point)? If it does, chose the one that maximizes profits. If it doesn't, no optimal integer solution exists and the airline must either relax some of the constraints or make do with a suboptimal solution.

Geometry of numbers can also be applied to packing problems. In 1893 Minkowski announced in a letter to a colleague that the density of a lattice packing of spheres (a geometrical problem) is related to the so-called Zeta function (a concept from number theory). Twelve years later, in 1905, he published a proof of this assertion. He showed that, in fact, the best lattice packing in two dimensions must have a density of *at least* ½(1 + ¼ + ⅙ + ⅟₁₆ + ⅟₂₅ + . . .). This equals 0.8224 . . . and is quite a way off from 0.9068, the density of the hexagonal packing. But since, in 1905, it was not yet known whether the hexagonal packing is, in fact, the densest packing possible, this lower bound was considered quite an advance.[5] Unfortunately Minkowski's theorem is nonconstructive, which means that it only guarantees that such a lattice exists, but doesn't show what it actually looks like.

In 1892 Thue need not have worried that his German competitor would beat him to the finish line. Minkowski was still far from his 1905 proof and, anyway, his approach only dealt with the lattice version of the problem. But

[5] Minkowski's lower bound is still the best estimate for high dimensional spaces (see chapter 9).

Thue feared that Minkowski would, sooner rather than later, put his mind to the general version and possibly succeed. So he was in a great hurry to come out of the closet and announced a lecture on the subject. In that way he could always claim priority over Minkowski if the question ever arose.

The audience at the meeting in Copenhagen waited with bated breath to hear Thue, but was treated to a rather taciturn presentation. At least this is what we have to gather from the published report. The printed version of his sibylline talk contains all of twenty-three lines (plus one figure). But it was not its brevity that raised criticism. Thue only presented an outline of how the packing problem in two dimensions could be tackled. He may have had a good idea, but a proof it certainly was not. A well-known number theorist, Carl Ludwig Siegel, described Thue's attempt as "reasonable, but full of holes." An expert, Claude Ambrose Rogers, would write seventy years later that "the published account of the lecture is very short and is hardly sufficient to enable his proof to be reconstructed." In particular he criticized that "there is no [discussion of the] situation when seven of his points lie within a [critical] distance of one of the points." There will be more to say about the problem of the seven points later.

If Thue had meant his address to be a preliminary report on a possible strategy for a proof, the mathematical community waited in vain for a final version that might have filled in the holes. It never appeared.

Eighteen years later, in 1910, Thue tried his hand again, this time using a totally different method. He did not do so because he had become aware of any errors in his first paper—indeed, in a footnote he proudly mentioned his previous essay—but apparently felt that one more essay on circle packings would be in order. This time he had the good sense to write in German, which had long before replaced Latin as the *lingua franca* among mathematicians. Now at least his peers could also read it. They did, and the verdict was not long in coming: Thue's newest work did not fulfill the stringent requirements of the professionals, either.[6] Maybe it hadn't been such a good idea to write the paper in German. Had Thue published in Danish or Norwegian again, it would have been less open to scrutiny.

So neither of Thue's two attempts managed to convince later mathematicians. The community had to wait another thirty years for a satisfactory proof of the general circle packing problem. In 1940, the Hungarian mathematician Laszlo Fejes-Tóth appeared on the scene. Born in Szeged, in

---

[6] As a reviewer noted, the paper is "open to the objection that it is no easy task to establish certain compactness results which [Thue] takes for granted."

1915, he studied mathematics and physics at the University of Budapest. He got his doctorate in mathematics in 1938. A two-year service in the Hungarian army followed, after which he received his first academic post at the University of Kolozsvar. Positions in Budapest, Veszprem, and, again, Budapest followed. For fifteen years, from 1970 to 1985, he served as director of the Mathematical Institute of the Hungarian Academy of Science. Fejes-Tóth published about one hundred and eighty papers and two books in the areas of convex and discrete geometry. Even though he never had a single Ph.D. student his work and open problems have influenced almost every contemporary discrete geometer. He will also accompany us in some of the following chapters.

There must be something about nonlattice circle packings that makes for short papers. The fact that Fejes-Tóth's proof was twice as long as Thue's first paper isn't saying much. It consisted of only forty-seven lines, plus a figure. But even if the proof wasn't very loquacious, it certainly was rigorous—no more handwaving here.[7] The paper also contains four footnotes, and one of them turns out to be of some significance because it deals with the notorious case of the seven points.

Mathematical papers are not exactly known for their catchy titles, but even among this sour fare, Fejes-Tóth certainly chose a winner. Written in German, the paper carried a title reminiscent of Thue's article: "Über einen geometrischen Satz" (About a geometrical theorem). It was guaranteed to arouse little interest. Superficially, the theorem does not seem to have anything to do with circle packings. However, it is surprising in its own right and we shall first describe what it says. Then we shall outline how Fejes-Tóth proved it, and in the course of the proof it will become clear why this theorem gives a definitive answer to the general (nonlattice) circle packing problem.

Let's illustrate the theorem. A farmer planted five hundred trees in a field that he had leased. To permit sunshine to enter the forest and to allow for sufficient irrigation, he left a distance of at least 2 meters, sometimes more, between any two trees. When the owner of the land died, his son presented the farmer with a choice: purchase the area on which the trees stand at a price of $20 per square meter or forget about forestry. The poor farmer had $34,000 at his disposal and had no idea how large his field was. He tried to make a quick mental calculation of how much he would have to pay, but, being a rather simple man, he just stood there scratching his head. Without

---

[7] Fejes-Tóth was kind enough to give Thue credit as having been the first to prove the packing problem in two dimensions, even though he was surely aware of that proof's shortcomings.

even consulting a surveyor Fejes–Tóth could have told him immediately that he did not have enough money: the area he would have to purchase could not be smaller than 1,732 square meters. How did Fejes–Tóth arrive at this number? To find out, read on.

Take an area of any size and shape, and place a very large number of dots into this region. (We denote the number of dots by N. In fact, the theorem holds if N is infinitely large.) You can place them any way you like. You could throw darts onto a board, drop the dots randomly into the area, or place plants in a pattern in a flowerbed. The theorem that we are discussing holds for any kind of arrangement of dots, darts, and flowers. The only condition is that none lie on top of another. Once the dots have been placed, take the two that lie closest to each other and measure the distance between them, denoted by D. Now place squares of side length D around each of the dots. There may be overlaps and there may be gaps, but that does not bother us. We have now packed or covered the area with N squares, each of which covers an area of D^2. The total area covered by the squares—counting overlaps twice—is ND^2. Fejes-Tóth's theorem states that the total area of the squares cannot surpass the surface of the original area by more than 15.5 percent. This is quite surprising. Why shouldn't there be more overlaps? Why can the squares not cover areas that are, say, 20 percent greater than the original area?

Let's try to fool the theorem, and increase the distance between the two dots somewhat. As soon as D becomes large enough, ND^2 will become greater than the original area plus 15.5 percent, right? Wrong! When the distance between two neighbors is sufficiently increased, other dots will become too crowded. Suddenly, two other dots will be identified, such that the distance between them, let's denote it by F, fulfills the theorem. In other words, as soon as ND^2 surpasses the original area by more than 15.5 percent, two other dots will be close enough, such that NF^2 fulfills the theorem.

Where does the number 15.5 percent come from and how can the theorem be proven? To replicate Fejes-Tóth's proof, first recall that all dots lie at a distance of at least D from each other. Now consider three dots that lie as close to each other as is allowed; they lie at the corners of an equilateral triangle with side length D. In what follows we shall need a very special number—the radius of the circle that cuts through the three corners of that triangle (there is only one). It equals $0.577 \cdot D$ (see appendix). We now place circles with this radius around each of the dots. This guarantees that no part of the area will be covered by more than two circles.

Before we continue with the proof we must answer the following question. We know that not more than two circles can overlap at any point, but how

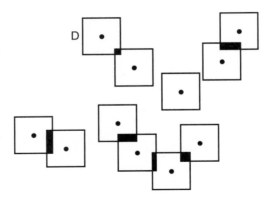

Dots in an area

many circles can reach into another circle? The answer is at most, seven. Why can't eight or more circles reach into a central circle? In order for circles with radii R to overlap—or at least touch—their centers cannot lie farther away from each other than twice their radius ($2R = 2 \cdot 0.577 \cdot D = 1.154 \cdot D$). On the other hand, the surrounding dots must lie at a distance of at least D from the central dot, because that was specified as the smallest distance between any two dots. This is the same as saying that the surrounding dots must lie outside a circle of radius D from the central dot.

Let's recapitulate. If we place one dot in the middle, then the centers of the overlappers must lie farther away than D and, simultaneously, not farther away than $1.154 \cdot D$. These two conditions define a ring-shaped area around the middle dot, inside which the surrounding dots must lie.

But the dots must also lie at a distance of at least D from each other. So how many dots can be placed inside this ring that satisfy this condition? Let's make

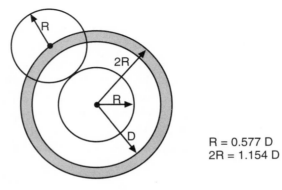

R = 0.577 D
2R = 1.154 D

Circles with radius R, D, $2R$

a feeble attempt at arranging eight dots in a judicious manner. The way to place as many dots as possible into the ring-shaped area—with the maximal distance between them—would be to arrange them on the outside border of the ring. The appendix shows that such an arrangement results in distances of $0.884 \cdot D$ between each pair of dots. This is too short, and eight overlappers are impossible.

What about seven overlappers? Let us again place the dots on the outer circle and check whether they have the required distance from each other. The appendix shows that seven equidistant dots on the outer circle result in a distance between any two neighbors of $1.002 \cdot D$. This is a hair's breadth larger than D, hence the requirement is satisfied, and seven overlappers are a real possibility.

We continue the proof by asking what area is covered by the circles? In order to compute this surface, care must be taken to count the areas that are covered by two circles only once. Fejes-Tóth used a neat trick to do that. He simply split the double-covered areas down the middle by drawing a straight line through the wedge-shaped sectors. One-half of the overlapped area is allocated to one circle, the other half to the other circle. Apart from avoiding double counting, this trick has the additional advantage of simplifying the mathematics. Instead of awkwardly dealing with the surfaces of circles and wedges, one needs to compute only the areas of straight-edged figures, which are easier to handle.

So what is the total area that is covered by the circles? Let us inspect one typical circle that is surrounded by other circles. Obviously, the covered area is smaller if the central circle is partially overlapped by one of the outside circles. The area is smaller still if the central circle is overlapped by two outside circles. And the greater the number of outside circles that overlap the central circle, the smaller the area gets, right? Wrong again! Let's see why. The key to the

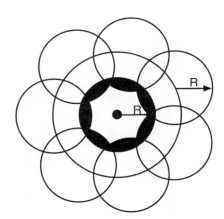

Circles surround a central circle

Allocation of wedges

problem lies in the overlapped wedges, since they are the areas that save space. If the overlapped area is big, then a big part of the surface is allocated to the outside circles, and the remaining area—the surface allocated to the central circle—becomes smaller. So let's try the whole thing again.

If one outside circle reaches into the central circle we get one wedge, half of which is allocated to the outside circle. Let a second circle reach into the central circle. There are now two wedges, and half of their area is allocated to the outside circles. And if three circles reach into the central circle, the overlapped areas are three times as large. And so on it goes, the area allocated to the central circle becoming smaller and smaller until circle number six. But at that point something happens: there is not enough room to add circle number 7! In order to accommodate the seventh circle in their midst, circles 1 to 6 must make room and move toward the outside edge of the ring. They must move so far toward the outside circumference of the ring that they hardly reach into the central circle anymore. As a consequence, the seven wedge-shaped sectors are so minute that the area that remains allocated to the central circle is larger than it was with only six overlappers. So the smallest area is allocated to the central circle when six circles reach into it.

Until now the proof dealt with a typical circle in the center of the plane, surrounded by a number of outside circles. But "center" and "outside" are relative terms when the number of circles, N, goes to infinity. Each circle can be considered, in turn, a central circle or an outside circle. And since half of the surface

  ....

1, 2, 3, . . . , 6, 7 circles surrounding a central circle

of each wedge was allocated to each of the two circles, the whole argument can be extended without further ado to all N circles in the plane.

We have thus shown that, no matter how the dots are arranged, the area covered by the circles can never become smaller than the configuration where six circles reach into the central circle. And in order to overlap in the most efficient way, that is, in such a way that the total surface of the six wedges is at a maximum, the six outside circles must be placed in a hexagonal arrangement. How much of the area is then covered? If each wedge is cut through the middle, the area allocated to the central circle is a straight-edged hexagon with side length D whose surface is easily determined. The appendix shows that this hexagon covers an area of $0.866 \cdot D^2$. (The number 0.866 is equal to $1/(1+15.5\%)$.) In other words, each dot requires a surrounding area of at least that surface. Fejes-Tóth thus showed that the total surface, call it T, that is needed to position N dots at a distance of at least D from each other, must be greater than $0.866 \cdot ND^2$. Turning the argument around, we can say that ND^2 must be smaller than $1/0.866$ times the area T. This is equal to T plus 15.5%, which is what we set out to prove.[8]

❖ ❖ ❖

This is why the farmer's land must be greater than 1,732 square meters: 0.866 times 500 trees, times 2^2. What does all that mean for circle packings? Fejes-Tóth showed that the surface must be greater than a certain lower limit—no matter how the dots are arranged. This limit is attained when the dots are placed in such a manner that six circles reach into a central circle. Hence, the closest packing of circles is achieved when the dots, which represent the center of the circles, are placed in a hexagonal pattern. As we saw in the previous chapter, this is the same result that can be deduced from Lagrange's findings, obtained in 1773. But Lagrange's result was achieved under the assumption that a lattice arrangement underlies the placement of the circles. Fejes-Tóth did not place any restriction on the placing of the dots. Therefore, his demonstration is a proof for the *general problem* of packing circles as closely as possible.

To the casual reader it may seem that Fejes-Tóth simply pulled a hexagon out of his hat, but the argument is sophisticated. The crucial insight was provided by the fact that, in order to save space, exactly six circles must reach into the central circle. The hexagonal arrangement was a consequence of the attempt to maximize the area of the wedges. One significant advance of Fejes-Tóth's proof over the previous attempts was his footnote number 4, in which he treated the problem case of seven circles, which had been completely neglected by Axel Thue.

[8] In mathematical notation this reads as $T \geq 0.866\, ND^2$ or $ND^2 \leq 1.155\, T$.

As if the locks had been opened, Fejes-Tóth's proof made way for more work on the two-dimensional packing problem. There was no hysteric rush, mind you, but in 1944, four years after Fejes-Tóth's paper had appeared in the *Mathematische Zeitschrift,* two mathematicians from the University of Manchester in England, Beniamino Segré and Kurt Mahler, published the paper "On the Densest Packing of Circles" in the *American Mathematical Monthly.* Now here was a title that at least made sense. Moreover, the paper was written in English. While they were working on the problem the authors were unaware of Fejes-Tóth's earlier result. Only when a colleague, Richard Rado of Sheffield, read their manuscript, did he draw their attention to Fejes-Tóth's previous work. Rado also informed them that he too was working on the problem, but he probably became discouraged by the deluge of proofs and never published his findings.

Segré and Mahler met at the University of Manchester. The two men came from totally different backgrounds but they had three things in common: their year of birth (1903), an interest in circle packings, and their faith: they were both Jewish. Their religious denomination was of great significance in Europe in the 1930s and the consequences that arose were what brought them together in Manchester. Segré was born in Turin, Italy (the same town from which Joseph-Louis Lagrange hailed), and Mahler in the German town of Krefeld. Segré came from a family of well-known scientists and was a gifted student. He received his doctorate at age twenty, and posts in Turin, Paris, and Rome followed. At the age of twenty-eight, when he was appointed to a chair in Bologna, he already had forty papers in various branches of mathematics to his credit.

Mahler, on the other hand, had a less fortunate youth. From early childhood he suffered from tuberculosis and as a consequence had to leave school at the age of thirteen. He took a job as a factory worker but never stopped pursuing his studies. After work he taught himself mathematics, and even tried his hand at mathematical research. His proud father sent some of the small articles to a mathematician he knew, who passed them on to friends of his. They finally ended up in the hands of Carl Ludwig Siegel, a professor at the University of Frankfurt. Siegel was suitably impressed with what he saw and arranged for Mahler's admission to the Universities of Frankfurt, and later Göttingen. In 1927 Kurt became *"Herr Doktor* Mahler," but just when his academic career was about to take off, with an appointment at the University of Königsberg, the Nazis came to power in Germany. At the same time, the Fascists took over in Italy.

Both Segré and Mahler were affected to the extreme by the political developments because of their Jewish backgrounds. The Fascist Italian government forced Segré out of the University of Bologna, and Mahler decided to leave before the Nazis would kick him out of the University of Königsberg. They departed from their respective homelands and made their ways to the shores of England. They were not made to feel especially welcome. As was the unfortunate custom at the time, after entry to Britain every person who hailed from Germany or Italy—Jewish or otherwise—was first interned as an "enemy alien." Segré and Mahler were no exception and they got to enjoy the dubious honor of being hosted by His Majesty's detention service for a few months in 1940.[9] After they were set free, they met up at the University of Manchester, at the invitation of the mathematician Louis Mordell. Segré and Mahler were forty-one years old when they collaborated on their paper on circle packings.

They approached the problem differently from Fejes-Tóth. They asked how many circles of radius 1 can be packed into a triangle, a square, a pentagon, or a hexagon of area T. Their answer: at most $0.289 \cdot T$. To prove this assertion, Segré and Mahler surrounded each circle by a straight-edged cell, and then computed the surface of these so-called *Voronoi cells*.[10]

Let me illustrate the situation with the following scenario. A group of investors purchase an oceanfront property, of size T, in order to build Voronoi Village, a holiday retreat. The developers want to parcel out the property into separate lots. Inspired by Buckminster Fuller's *Dymaxion* houses, each lot would contain a round villa propped on a pole (that's how Bucky designed them) and surrounded by a little fenced garden. The property boundaries would run down the middle of the space between any two villas, as indicated by the fences. These boundary lines would, when viewed from above a given villa, form polygons with as many straight edges are there are surrounding villas. If two villas were adjacent, the garden fence would simply run through the point of contact. Like all property developers worth their salt, the Voronoi Village Consortium wants to maximize profits by squeezing as many villas onto the oceanfront property as possible. How many can they fit?

Segré and Mahler set out to calculate the areas of the separate lots. For the sake of the calculations they assumed that the villas have a radius of 1. They partitioned the lots into triangles and showed that each of these triangles must be greater than 0.5513 multiplied by the appropriate angle in the middle of the villa.[11] (The details of the calculations are quite intricate,

[9] By the way, Richard Rado was born in 1906 in Berlin and, like Segré and Mahler, was Jewish. His family sensed the coming catastrophe and moved to England in 1936. Thus, unlike Segré and Mahler, Rado was spared the internment.

[10] In chapter 9 I will have much more to say about Voronoi cells.

[11] Angles are measured in radians, between zero and 2π.

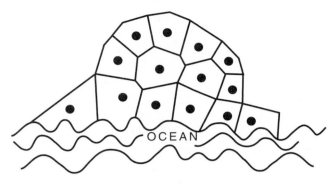

Voronoi Village

and I won't repeat them here.) Since the angles of all the triangles of a lot must add up to a full circle, which is equal to 2π (in radians), the sum of the triangles' surfaces is greater than 0.5513 times 2π. This equals 3.464, and it means that each lot covers an area of at least 3.464. Segré and Mahler concluded that Voronoi Village, the oceanfront property of area T, cannot contain more than $T/3.464$ separate lots.

Now I will translate the Voronoi Village Consortium's experience back into language mathematicians can understand, and see what all this means for circle packings. Segré and Mahler's calculations imply that no greater density can be attained in two dimensions than when $T/3.464$ circles are packed into a polygon of surface T. And which arrangement achieves that density? The answer is—surprise, surprise—the hexagonal arrangement! So it has been proved once again that the configuration of "six around one" is the densest packing of circles in the plane.[12]

And what's the density? Recall that density is defined as the ratio of the area covered by the circles divided by the total area. Since a circle of radius 1 covers an area of π, and since there are $T/3.464$ circles in a polygon of area T, we get a density of $\pi \cdot (T/3.464)/T$, which equals 90.69 percent. Does that look familiar? It better, since this number is the maximal density of spheres in two dimensions, as already pointed out in chapter 3.

Now that the packing problem has been solved a couple of times over—even in the general, nonlattice version—let us pick up a thread we left hanging in the previous chapter. I refer to *coverings* in the plane. Recall that

[12] In fact, Segré and Mahler showed that triangles, squares, or hexagons of area T contain *exactly* $T/3.464$ circles of radius 1, *iff* (mathematical shorthand for "if and only if") every circle is touched by six other circles. By the way, Segré and Mahler also alluded to, and correctly dismissed, the problem case of "seven around one."

a covering is defined as a set of circles that completely covers the plane, even if some parts of the plane are covered by more than one circle.

We go back to the late 1930s. On the other side of the Atlantic, at the University of Wisconsin, the mathematician Richard B. Kershner was working on circle coverings. Kershner was born in Crestline, Ohio, in 1913. When he was one year old, his father was appointed headmaster of the Franklin Day School and the family moved to Baltimore. At fourteen, Kershner went to sea as a deck boy on a freighter. Two years later, the sixteen year old entered the engineering program at Johns Hopkins University. But adventure called, and after his first year in college he again went to sea, this time as an ordinary seaman. Back from this exploit Kershner was admitted to a graduate program at Hopkins for qualified undergraduates. He received his Ph.D. in mathematics in 1937. By that time he had already published more than a dozen articles in the *American Journal of Mathematics.*

After graduation, Kershner was appointed instructor of mathematics at the University of Wisconsin. He stayed there for three years. When he returned to the mathematics department of his alma mater as an assistant professor of mathematics, he no longer had the leisure to concern himself with pure mathematics. Times were hectic and the atmosphere was charged: the United States was about to enter World War II. Everybody and his uncle were enlisted in the war effort. The young mathematician turned to ballistics, as always a thankful subject for mathematicians involved with the military. (Remember Lagrange and the shooters in Turin?) During the following years he and a colleague established the fundamental understanding necessary for the development of advanced ordnance and rocket systems. Thus Kershner did his part to defeat the Fascists and the Nazis who had thrown Segré and Mahler out of their respective homelands.

As dark clouds gathered over Europe, Kershner was busy with the covering problem. He wrote a paper while at the University of Wisconsin with another sensible title: "The Number of Circles Covering a Set." It was submitted to the *American Journal of Mathematics* in December 1938, and published in the following year. Kershner thus solved the covering problem a year before Fejes-Tóth solved the packing problem.

Let us cover a surface with circles and connect the centers of all circles with the centers of neighboring circles. A net of triangles results. There is a very neat result by Leonhard Euler about such nets. Count the number of faces (*F*), edges (*E*), and corners (*C*). Now add the number of faces to the number of corners. Euler showed that this sum is always equal to the number of edges plus 1. No matter what kind of net you think of, you will always get $C + F = E + 1$.

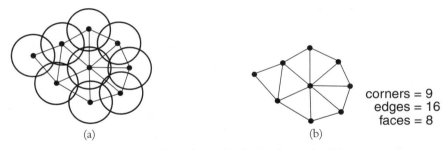

corners = 9
edges = 16
faces = 8

Covering (a), and associated net of triangles (b)

It is quite easy to prove this surprising theorem.[13] Start with one triangle. It has three corners, one face, and three edges. Hence Euler's theorem is satisfied: $3 + 1 = 3 + 1$. Now add two edges to form a net consisting of two triangles. Again Euler's theorem is satisfied: four corners, two faces, and five edges: $4 + 2 = 5 + 1$. Sometimes it may be possible to form a net of three triangles by just adding a single edge. Euler's theorem is still satisfied, because even though the extra edge added an extra face, no new corners were added: $4 + 3 = 6 + 1$. A third possibility of adding faces to the net is as follows. Weave an additional knot into the middle of an existing triangle by connecting that knot with the three existing corners. In this case one new corner is added, two new faces (*one* triangle was converted into *three* smaller triangles), and three new edges. We get $5 + 5 = 9 + 1$. Continue with this game as long as you like, the result always stays $C + F = E + 1$. The underlying reason for this is that in order to add a new triangle at the border of the net you must either add two edges, in which case you simultaneously increase the number of corners by one, or you just add one edge, in which case no additional corners appear. If you add a corner in the middle of the net you willy-nilly add two faces and three edges. Euler's theorem is satisfied in each of these cases. We will make use of this result in the following paragraphs.

I will now describe Kershner's proof. How large can the triangles of the net be? Recall first that the corners of the net represent the centers of circles that cover the net. Now draw *disks* around the triangles. (To avoid confusion with the original circles whose centers form the corners of the net, we call the circles around the triangles *disks*.) The distance from the center of the triangle to its corners cannot be larger than 1, so these disks have a radius of less than 1. (If the disks had a larger radius the circles would not overlap.) How large can a tri-

---

[13] Actually, the net need not even be composed of triangles. Any combination of polygons—combined into a net—satisfies Euler's theorem.

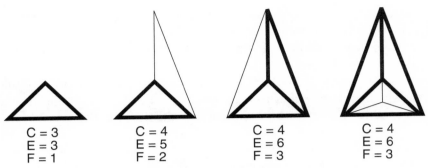

C = 3
E = 3
F = 1

C = 4
E = 5
F = 2

C = 4
E = 6
F = 3

C = 4
E = 6
F = 3

Euler's theorem, showing corners (C), edges (E), and faces (F)

angle that is inscribed into a disk of radius 1 be? This question is a different version of Queen Dido's problem (discussed in the previous chapter): Which triangle, inscribed into a disk of radius one, has the greatest surface? The answer is the equilateral triangle. In the appendix to this chapter we show that such a triangle has an area of 1.299. Hence each triangle in the net must be smaller than that number. Since the net is composed of F triangles, one for each face, the net's total surface must be smaller than $1.299 \cdot F$.

Our next task is to find a relationship between the number of faces in a net and the number of corners. Each face of the net is surrounded by three edges. Hence there are three times as many edges as there are triangles. But this would be double counting, since most edges belong to two triangles, one on each side of the edge. Let's call these the interior edges. Only the edges at the border of the net belong to a single triangle. If each triangle is allocated half of the interior edges, and all of the border edges, then the average number of edges per triangle must be higher than 1½, that is, $E \geq 1\frac{1}{2} F$. On the other hand, Euler

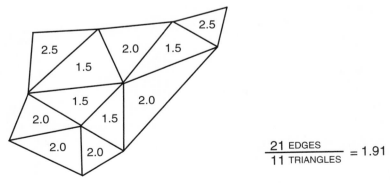

$$\frac{21 \text{ EDGES}}{11 \text{ TRIANGLES}} = 1.91$$

Average number of edges per triangle

stated that $E = C + F - 1$. Combining these two facts, we can give an upper limit for the number of faces of a net: $F \leq 2C$, that is, the number of faces is less than (or equal to) twice the number of corners. And this means that the total area of the net is smaller than $2C \cdot 1.299 = 2.598 \ C$.

Let's compute the density of the area covered by the circles. There are C corners in the net and each of them represents a circle. Since the radius is 1, each of these circles covers an area of π. The net's total area is smaller than $2.598 \ C$, as we just showed. Hence the density—the area covered by the circles, divided by the area of the net—is at least . Substituting the numerical value for π we get 1.209. This is sort of a magic number. Even if the net were to be made infinitely large, by adding more and more triangles, the density could never drop below 1.209. But if the number of triangles increases how can the density stay constant? Well, the number of corners, C, appears both in the numerator and in the denominator of the density equation. Hence it cancels out of the fraction, and so it does not matter how large the net is or how many triangles there are. The density of any net can never be smaller than 1.209. There will always be unavoidable overlaps to the tune of 20.9 percent.

One last question: Which circle arrangement attains the lowest density? Yes, it's the good old hexagon. The key to realizing this is that the only triangle in the unit circle that *exactly* covers an area of 1.299 is the equilateral triangle. And equilateral triangles placed next to each other make up a net of hexagons.

❖ ❖ ❖

Time out for an illustration. The manager of a beachfront has just received an offer for new parasols: top quality, low price, great colors, and a radius of 1 meter. His beachfront measures 26 meters by 10, and he wants to have all of it in the shade. Recently a high-school student joined his operation to make some money watching the action and arranging parasols. Having just completed an honors course in geometry, he proudly informs the manager that the shade of each parasol covers an area of 3.141 square meters.[14] The manager wants to put this handy information to good use. He pulls out his pocket calculator, computes that the area of his beach is 260 meters, divides this number by 3.141, and places an order for eighty-three parasols. But the parasol company had put its sales reps through an intensive course on "Applications of advanced geometry to the placement of parasols, with special emphasis on Euler's theorem." The sales rep tells the manager point blank, "Forget it! At best, each parasol can provide shade for no more than 2.598 square meters. Even if you place the parasols in the

---

[14] $3.141 = \pi r^2$. This is at noon, when the sun is at its zenith.

best method possible, you'll need at least one hundred of them to provide sufficient shade for the whole beachfront." Needless to say, the manager didn't believe one word the rep was saying and went ahead with his order of eighty-three parasols. After they arrived, he told his assistant to stick them into the sand. The poor boy kept arranging and rearranging them, but try as he might, there were always patches of sun shining in between the parasols. After numerous complaints from the patrons, the manager placed an order for another seventeen parasols, and the assistant went back to school.

So the hexagonal arrangement provides the best packing and, with somewhat larger circles, also provides the best covering. This completes the proof or, as mathematicians are fond of saying, *quod erat demonstrandum*. This Latin expression means "this is what had to be shown," and is usually abbreviated as QED. With a sigh of relief, this acronym is usually placed on the right margin of the page after the last line of a proof. Like so:

$$QED$$

Those readers who had sufficient courage to walk through this proof may have sensed some of the magic and beauty of mathematics. A number of steps were taken, each of which seemed quite plausible by itself, maybe even a bit trivial. But at the end of the path you realize that something remarkable has happened: a surprising, by no means trivial statement has been proved. In fact, it may have taken many mathematicians many years of intense effort to arrive at the statement's proof.

With this we have completed our discussion of Kepler's problem in two dimensions. We started with the lattice version of the packing problem in the previous chapter. In this chapter we first continued on to the general version of the packing problem. Now we have also shown how the general version of the covering problem was solved. (No need to prove the more restrictive, lattice version of the covering problem. The general version, as proved by Kershner, subsumes it.) In the next chapters we leave the flatlands in order to, once again, enter the real world of three-dimensional space.

CHAPTER 5

Twelve's Company, Thirteen's a Crowd

In this chapter we move back in time to the end of the seventeenth century and up in space to three dimensions. In 1694, a famous discussion between two of the leading scientists of the day—Isaac Newton and David Gregory—took place on the campus of Cambridge University in England. Their dispute concerned the "kissing problem." But don't get your hopes up. The term *kissing* in this context has nothing to do with the gesture of affection. Here the verb *kiss* refers to the game of billiards, where it signifies two balls that just touch each other.

Newton and Gregory argued about the number of spheres of the same radius that could be brought into contact with a central ball. On a straight line two balls can kiss a ball in the center, one on the left and one on the right. On a billiard table, at most six balls can touch a central ball. There's no room for a seventh, as anyone can verify by rolling the balls around a bit. The reason for this is that a heptagon (seven-cornered polygon) whose sides have length 2 (i.e., the diameter of a ball) is too large to fit tightly around a circle with radius 1. So far everything is clear. But let's move off the green felt and up into the realm of space.

In the 1950s, H. W. Turnbull, an English school inspector, was doing research on the life of Isaac Newton. Working his way through the numerous papers, letters, and notes, he came across two documents that would provide the basis for the kissing problem: a memorandum of a discussion that the two scientists had at Cambridge, and an unpublished notebook at Christ Church at Oxford in which Gregory had jotted down some notes.[1]

---

[1] Christ Church is the cathedral of the Anglican diocese of Oxford and also a college within the University of Oxford.

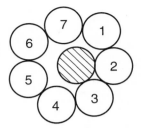

Heptagon arrangement: no touching
and no kissing

The two men had been discussing the distribution of stars of various magnitudes that revolve around a central sun. In the course of their deliberations the question arose whether a sphere can be brought into contact with thirteen others of the same size. And that's where opinions diverged.

How many white billiard balls can kiss a black billiard ball in three-dimensional space? Kepler stated in his treatise on six-cornered snow that twelve spheres could touch a central sphere, and then went on to describe two possible ways to arrange the balls. But maybe thirteen spheres can be brought into contact with a central sphere. Initially we may think this clearly impossible, since Kepler's arrangement is completely rigid. All the balls touch the central ball and they also touch each other. So how could a further ball be squeezed in between? The question is not quite trivial because the hexagonal packing—three balls below the central sphere, six around it, and three above—is not the only arrangement of "12 around 1." We already saw that the cubic packing—four balls below the central sphere, four on the sides, and four above—is another display of "12 around 1." As Kepler pointed out, however, these two arrangements are identical, and any apparent differences are merely the result of looking at the balls from different perspectives. But maybe there is a truly different arrangement that brings a dozen white balls into contact with the black ball.

There is—and not just one. Put one sphere on the bottom, then arrange five balls in a pentagon around the central ball, just below its equator, place another five balls more or less in the interstices of the lower five balls (this puts them slightly above the equator of the central ball), and finally top it off with the twelfth sphere on the pinnacle. So there you have it: another arrangement of a dozen spheres around the central ball. You may note that the spheres sit more or less on the vertices of an icosahedron, which is why this configuration is called the icosahedral arrangement.

If we now take a close look, a very surprising fact emerges: this arrangement is not rigid. Sufficient space is left over in the interstices between the twelve balls that they can roll around a bit on the surface of the central ball. You may ask yourself, as did Gregory: Can the balls be moved in such a way

Top
↓

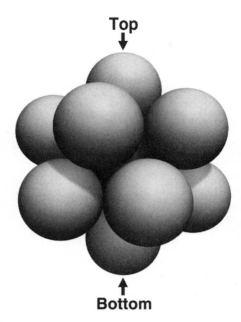

Icosahedral arrangement   ↑
                                        Bottom

that sufficient space opens up for an additional sphere? Maybe the free spaces can be combined so that a thirteenth ball can be squeezed in. This may seem absurd, but Gregory did have a point and we will give a mathematical demonstration that "13 around 1" is at least conceivable.

Consider a number—as yet unknown—of spheres that kiss a central sphere, all of radius 1. Deposit the whole arrangement into a superball with a radius of 3. Imagine a lamp at the center of the central ball that casts shadows of the surrounding balls onto the inside surface of the superball. These circular shadows cannot overlap. It is shown in the appendix that each shadow has a surface area of 7.6, and the total surface of the superball is 113.1. So how many shadows can fit onto the superball's surface? Divide 113.1 by 7.6 and you get 14.9. The inescapable conclusion is that there is room for nearly fifteen balls! Definitely there is sufficient surface, at least theoretically, for fourteen balls, and so thirteen balls certainly should be considered a possibility.

All of a sudden Gregory's claim does not sound as absurd as it did at the outset. You may nevertheless think it somewhat strange that he steadfastly claimed that an arrangement of "13 around 1" exists, but never came up with the recipe of how and where to place the balls. Why did he not simply position 13 balls around a central sphere and show his arrangement to Newton? *That* would have convinced him. Well, it was very difficult to

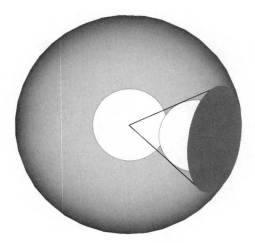

Superball

produce perfectly round spheres of wood or stone with the tools of the time, so experiments were never quite satisfactory. As a consequence, the question stayed around for another two and a half centuries. In fact, the dispute was only settled in 1953.

Let us digress for a moment. The Englishman H. S. M. Coxeter[2] from the University of Toronto showed in 1963 that Kepler's arrangement of "12 around 1" (the hexagonal arrangement) can be transformed into the icosahedral arrangement just by rolling the balls around, without lifting them off the surface of the central sphere. Since the icosahedral arrangement also allows the balls to roll around, one ball can be pushed a bit in this direction, another one in that direction, and so on. Some years later, the designer, inventor, and architect Buckminster Fuller named the process that converts the hexagonal arrangement into the icosahedral arrangement the *jitterbug transformation*. (There will be much more to say about Bucky Fuller in chapter 10.) This has far-reaching consequences. It means that in between the hexagonal arrangement and the icosahedral arrangement there exist an *infinite* number of configurations of "12 around 1." And that's another reason why the kissing question stayed around for so long: among the infinite number of arrangements, there just might be one that opens up sufficient space for a thirteenth sphere.

---

[2] The initials stand for Harold Scott MacDonald. Originally he was to have been named Harold MacDonald Scott but friends pointed out to his father that the name H. M. S. Coxeter would make the boy sound too much like one of Her Majesty's Ships.

The space left over in the interstices of the icosahedral arrangement occupied the minds of generations of mathematicians. But it must be pointed out that even after the Newton–Gregory dispute was eventually decided, Kepler's conjecture could not be put to rest. Just because a thirteenth sphere can, or cannot, touch a central sphere does not settle the question of the densest packing. The number of spheres surrounding *one* central sphere says nothing about how balls may be arranged in infinite space. To put it in mathematical terms, the kissing problem deals with a *local* question, while Kepler's conjecture is a *global* problem.

Let us get back to the dramatis personae. Isaac Newton was one of the foremost scientists of all times. If Johannes Kepler was considered the prince of astronomy, then Isaac Newton was surely the prince of physics. He was born on January 4, 1643, in Woolsthorpe, in Lincolnshire, a tiny, weak baby that was not expected to survive for even a week. However, Newton proved doctors and midwives wrong and extended his lifetime another eighty-four years.[3] He never knew his father, an uneducated, illiterate man who died three months before his birth. After the death of her husband, Isaac's distraught mother, Hannah, née Ayscough, was badly in need of the church's compassion, and the minister of the nearby village, who went by the name of Barnabas Smith, was only too happy to oblige. The reverend took his comforting duties very seriously and after a proper period of mourning the two got married. With this, little Isaac became superfluous and was shipped off to his grandparents, who weren't too pleased with the sudden appearance of a two-year-old boy. Isaac did not have a happy childhood with them. There was no love lost between him and his grandfather, James Ayscough. The old man even excluded him from his will. Isaac was furious. About his mother and stepfather he fantasized that he would "burn [them] and the house over them."

Newton's aim at college was to get—what else—a law degree. At Cambridge, that included studying the antiquated texts of Aristotle, which still put the earth at the center of the universe and described nature in qualitative instead of quantitative terms. But revolutionary ideas can't be repressed—they just float around at centers of learning—and sometime during his undergraduate days Newton discovered René Descartes' natural philosophy. Descartes viewed the world around him as particles of matter and explained natural phenomena through their motion and mechanical interactions. Through Descartes' writings Newton was led to study—without anyone's knowledge—the important mathematical texts of the time. Then his genius began

---

[3] Some sources put Newton's birth on Christmas Day of the previous year, but that's because the Gregorian calendar was adopted in England only a hundred years later.

Isaac Newton

to emerge. All by himself he created revolutionary advances in mathematics, optics, and astronomy. Single-handedly he invented a new mathematical method to describe motion and forces, which he called the "method of fluxions," eventually known as calculus. To his greatest regret later in life, he never published an account of the method that allowed the computation of areas, lengths of curves, tangents, and maxima and minima of functions. He probably thought that the new technique was too radical a departure from traditional mathematics, and that discussions about it would detract the readers of his astronomy texts from the main results. Therefore, he may have deliberately covered his tracks by keeping the invention of differential calculus a secret. The treatise in which he eventually described the new method, "De methodis Serierum et Fluxionum" (On the methods of series and fluxions), written in 1671, would only be published sixty-five years later, ten years after his death.

Newton's greatest discovery was the universal law of gravitation. Johannes Kepler had discovered that planets moved in elliptic orbits. Then there had been the apple falling from the tree. These clues, paired with phenomenal insight, allowed Newton to establish the fact that the force acting on a planet decreases with the square of its distance to the sun. Kepler had shown how, but Newton explained why. At the urging of his friend, the astronomer

Edmund Halley, Newton wrote *Philosophiae naturalis principia mathematica* (Mathematical principles of natural philosophy), which was published in 1687. In the *Principia,* which has been described as the "greatest scientific book ever written," Newton analyzed the motion of bodies under the action of centripetal forces and gave explanations for orbiting bodies, projectiles, pendulums, the free fall of bodies, the tides, eccentric orbits of comets, the precession of the earth's axis, and the motion of the moon.

But most of his colleagues would not accept the physical law of gravity. The law, which states that all matter attracts all other matter, baffled his contemporaries. Everybody could imagine pulling an obstinate donkey by a string, but pulling a donkey without a string? That was too much for the common public and even for the learned men to believe. Obviously a string is needed to connect the donkey to the yanking farmer, and clearly one body can move another body only when the two are in contact (when they *kiss,* so to say—which does, after all, denote a moving experience). An invisible force just could not act through thin air or through a vacuum, Newton's colleagues thought. They had only recently stopped believing that a magician could, just by waving his hands, put objects in motion at the other end of the room. It took a long time and much convincing until "action at a distance" was universally accepted.[4]

At age fifty-three, Newton decided on a career change. He became Warden of the Royal Mint, and three years later, until his death, he was its Master. Since he received a commission on all the coins that were struck, he managed to do quite well for himself. He also pursued counterfeiters relentlessly and with ferocity.

Many great people have a skeleton hidden in the closet and Newton is no exception. In Kepler's case it had been his infatuation with astrology; Newton's secret hobby was alchemy. He was obsessed with the idea of turning metal into gold. Unfortunately, the material he and many other alchemists used for experiments was mercury. As many of his colleagues discovered the hard way, not only does mercury not yield gold, it also has poisonous properties. The method that was generally used at that time to determine a compound's characteristics was to taste, smell, or swallow it. In the case of mercury, this wasn't such a good idea. Fortunately, Newton suffered no ill effects from his experiments.[5]

---

[4] Two and a half centuries later scientists had the same problem trying to explain Einstein's theory of relativity to the public.

[5] Some biographers believe, however, that Newton's depression in later life may have resulted from the inhalation of mercury.

In 1703 Newton was elected president of the Royal Society of London. He was reelected every year until his death twenty-four years later. Apart from dozing off during the meetings toward the end of his life, he used his presidency quite to his advantage, as we shall see. Sir Isaac devoted the last twenty-five years of his life to a priority dispute with Gottfried Wilhelm von Leibniz over the discovery of calculus—Newton's method of fluxions. The stakes were high. Calculus changed mathematics in a fundamental way, and its inventor would forever be remembered for this feat. As president of the Royal Society Newton appointed an "unprejudiced committee" to decide who had been the first to invent calculus. Of course the unprejudiced committee members knew exactly what was expected of them and didn't dream of letting Leibniz off the hook. The gentleman from Germany was never even asked for his version of the events. Just to be on the safe side, Newton secretly wrote the committee's final report himself. And to top it off, he also wrote a very favorable review of the report—again anonymously—for the *Transactions of the Royal Society*.

Nowadays it is generally accepted that Newton deserves priority, having invented calculus in 1665 and 1666. Leibniz apparently reinvented the method, which he called the "method of differences," ten years later. Isaac Newton died on March 31, 1727, in London.

The other participant in the debate on the kissing number was David Gregory, the nephew of the more famous scientist James Gregory. Newton's junior by sixteen years, Gregory was born in 1659 in Aberdeen, Scotland, to a very fertile family: he was one of his father's twenty-nine children (by two wives). A precocious child, he started his university education at the tender age of twelve. But the early start did not last. Rather it ended as quickly as it started and Gregory concluded his days as a student without a degree. This did not stop the University of Edinburgh from appointing him, at age twenty-four, professor of mathematics. Gregory was an early supporter of Newton. In fact, he was the first university lecturer to teach the cutting-edge theories that no other university had yet adopted. In 1690, when unrest set in in Scotland, he left for Oxford. Thankful for Gregory's endorsement of his theory, Newton arranged for his appointment as Savilian Professor of Astronomy.

The discussion on kissing numbers began when the fifty-one-year-old Newton was between jobs. He already had one foot in the Mint in London, but had not yet removed the other foot from the ivory tower of Trinity College. On May 4, 1694, Gregory paid him a visit. His stay at Cambridge lasted several days, during which the two men talked nonstop about scientific matters. It was a rather one-sided conversation, with Gregory, the dutiful disciple, making notes of everything the great master uttered. He had to hurry because Newton freely related his thoughts, jumping from one subject to another. From editorial corrections to the *Principia,* he went to the curvature of geometric objects, the "smoke" issuing from a comet, speeds of different colors of light, conic sections, the interaction between Saturn and Jupiter, and so on. All this happened at lightning speed, with Gregory attempting to take it all down. When he was not quite able to follow, Newton just took the pad from his friend's hands and scribbled his own remarks into the notebook.

One of the points discussed, number 13 in Gregory's memorandum of that day, was how many planets revolve around the sun. The discussion then went off on a tangent, to the question of how many spheres of equal size could rotate around a central ball of the same size. It deviated again to the distribution of stars of various magnitudes around a central sun. Finally Gregory asked the question: Can a rigid material sphere be brought into contact with thirteen other spheres of equal size?

In one of the notebooks found by H. W. Turnbull[6] there is a passage in which Gregory discusses the packing of circles that are placed in concentric rings around a central circle, that is, the two-dimensional problem. He correctly pointed out that six spheres can surround a central sphere in the innermost ring. This is the problem that was proved by Fejes-Tóth in 1940. He also remarked that the next rings contain twelve and eighteen spheres. Gregory then went on to discuss the question for three dimensions. How many spheres can be placed in concentric layers so that they all touch the central ball? It is here that he made a claim that sparked the debate and the 250-year search for a final answer. Gregory stated—without further ado—that in three-dimensional space the first layer surrounding a central ball contains *thirteen* spheres.

Newton, on the other hand, had written in *"A table of ye fixed Starrs for ye yeare 1671"* that there exist thirteen stars of the first magnitude, and in Gregory's report on the discussion of May 4, 1694, we read that in order "to discover how many stars there are of first, second, third etc. magnitude, [Newton] considers how many spheres, nearest, second from them, third etc. surround a sphere in space of three dimensions: there will be 13 of the first magnitude." Of course, Newton meant a total of thirteen spheres, including the central one.

So Gregory and Newton did not agree on the number of spheres that can kiss a central ball. If you were to place a bet, whose side would you take? Do you believe in "12 around 1" or in "13 around 1"? Betting on

---

[6] These notebooks are kept today at Christ Church at Oxford for safeguarding.

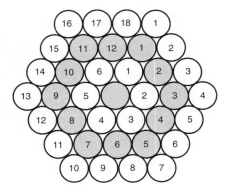

Rings containing 6, 12, and 18 circles

Newton would be a safe wager. While Gregory is considered a fair but by no means outstanding mathematician, Newton was correct in most things he ever said or did. But he was not infallible, as his attempts to make gold from mercury showed. And—in Gregory's favor—there is enough space for nearly fifteen balls, as we saw previously.

It turns out that Newton was right: "13 around 1" is impossible. That's why the highest number of balls that can touch a central ball is now often called the Newton number. But during their lifetimes the two men were never to know the correct answer, and after whom the kissing numbers would be named. And anyway, being correct is only half the fun in mathematics. The other half is finding the proof, and some headway on that track—though no definite progress—was only made 180 years after Newton and Gregory formulated their controversy. The final proof had to wait an additional eighty years and was only formulated in 1953.[7] In the next chapter I will describe the various attempts to solve the kissing problem.

[7] Readers may be forgiven if they do not find the arguments above totally convincing. After having finished the manuscript, this author remained beset by doubts. Does Newton's statement really indicate that he knew the correct number to be twelve? Or did he, like Gregory, think that thirteen spheres can touch the central sphere? Seeking an answer, I contacted Dr. Robert Hunt, Deputy Director of the Isaac Newton Institute for Mathematical Sciences at the University of Cambridge. We sought further sources to confirm whether Newton knew the correct answer. Alas, we were not able to find any, and until further notice the question remains open.

CHAPTER 6

Nets and Knots

In 1869 a mathematician by the name of Bender from the Swiss town of Basle submitted a proof for the kissing problem to the *Archiv der Mathematik und Physisk,* a well-regarded journal published in Germany. The paper was entitled "Bestimmung der grössten Anzahl gleich grosser Kugeln, welche sich auf eine Kugel von demselben Radius, wie die übrigen, auflegen lassen"—in English, "Determination of the largest number of equal-sized balls, that can be placed on a ball with the same radius, as the others." (There is nothing lost in the translation; even the original German title sounds awkward.) The problem's roots in Newton and Gregory's dispute were not mentioned and were apparently unknown to the author, who had just put his mind to a geometrical problem that he found interesting. Dr. Bender submitted his paper to the *Archiv* on May 25, 1869. And then he waited. And waited. And waited.

Altogether he waited for five years, until 1874, when his exposition finally appeared in print. There was a good reason for the delay, however, and it wasn't the confusing title. Bender's alleged proof was no proof at all. All he had managed to demonstrate was that twelve balls could touch the central sphere and that each ball touched four neighbors simultaneously. He did that by showing that if twelve balls are placed on the surface of the central ball in the familiar arrangement, then there remain eight empty triangles and six quadrangles in between the balls. Bender then computed the surfaces of these triangles and quadrangles, summed them, and concluded that there remains sufficient room for only the twelve balls with which he had started out. Toward the end of the paper he added, as an aside, that it is "easy to show that no other arrangement allows a greater [number of spheres]," and, therefore, "not more than twelve balls can be placed on the surface of a ball with the same radius."

These are big words indeed, but Bender had proved nothing at all. Worse still, he had committed the cardinal sin of establishing a tautology and trying to pass it off as a proof: if you start out with twelve balls, you end up with twelve balls. At best, Bender had shown that in the particular arrangement he considered, no thirteenth ball could be added to the twelve that are already there. That was no great achievement, since it was known all along—at least since Harriott and Kepler discussed the matter toward the end of the sixteenth century—that twelve balls could kiss a central ball. The real question was whether some *other* kind of arrangement exists that allows thirteen spheres to come into contact with the central ball. And that question had been left open by Bender.

Strangely enough, the editorial board of the *Archiv* felt that this nonsense was nevertheless worthy of publication. Why were the editors, typically concerned about the quality of the material that appears between the covers of their journal, so soft-hearted in this case? The reason for their leniency soon becomes apparent. Bender's alleged proof, which should never have seen the light of day, gave another mathematician by the name of Reinhold Hoppe the opportunity to present his own version of a solution to the kissing problem. Hoppe just happened to be the editor-in-chief of the *Archiv*.

Hoppe's entire life was exceptionally unspectacular and his publications record is no less mediocre. His research covered mathematics, physics, philosophy, and linguistics. In mathematics he published 250 articles, which seems an extremely respectable number at first sight. But 80 percent of the articles appeared in the journal that he himself edited, and many of them were simply "fillers," short notes that served to complete the sheet when another article ended in mid-page. The quality of his more weighty pieces was not outstanding either. Since he rarely bothered to determine what other authors (if any) had had to say about a given subject, much of Hoppe's original research was of only marginal interest. So insignificant were most of his papers that sometimes he completely forgot that he himself had written on the same subject a few years earlier. At the age of forty he stopped bothering with the pertinent literature altogether and henceforth was not even familiar with the names of the most famous of his contemporaries. A colleague described him as a "hermit of science." Hoppe had a very unpleasant appearance and his unpretentiousness and humility led to a very neglected demeanor.

But as editor of one of the foremost mathematics journals in Europe he wielded great power, which he could use as he liked. In his obituary, typically an occasion for summing up a person's accomplishments, a colleague wrote that Hoppe had composed many articles that would have better been left unwritten. His handling of submissions from outside contributors was equally casual. Especially toward the end of his life, his lack of intellectual astuteness allowed, "many conceited and loquacious authors, with little knowledge and less ability, to convince him to publish the mediocre and sometimes nonsensical products of their pens. But who would have had the heart to argue with an octogenarian?"[1]

So, to the problem at hand. In 1874, Hoppe decided to publish Bender's contribution but attached an addendum to the end of the faulty paper, entitled "Commentary of the editorial board." This was done rather discreetly. In fact, Hoppe didn't even sign the commentary. What probably happened was that when Bender submitted his paper, it was recognized by Hoppe as being deficient. However, it did manage to spark an idea that then took five years to gestate. Finally Hoppe was ready to present his proof to the world. But apparently he felt it would be unfair to publish his essay without giving Bender due credit. As a consequence, both papers were printed in the *Archiv,* one right behind the other.

Hoppe began his commentary by pointing out the shortcomings of the preceding paper. Starting from the arrangement that Bender considered, he noted that many movements and displacements of the balls were possible. The real question, he wrote, was whether the spheres could be pushed around in such a manner that sufficient room opens up to let a thirteenth ball squeeze in. This question was left unanswered by Bender, and now Hoppe set himself to it. I will only sketch the main points of his demonstration because a similar proof will be described in the appendix.

Assume that thirteen spheres *can* touch the central sphere, and consider the points where they contact the central sphere.[2] Let us weave a net around the central sphere. From each point of contact a thread is woven to its neighboring points of contact. If two threads cross, the longer one is deleted. The resulting net will have thirteen knots and consist entirely of triangles. The threads must be sufficiently long to accommodate two spheres next to each other, and from this Hoppe deduced that the net must consist of exactly twenty-two triangles. (He did that using Euler's theorem in three dimensions. Details follow.) Then he computed the surface of the smallest such triangle, multiplied that area by 22 (the number of triangles in the net), and concluded that even the smallest net will have too much slack to fit tightly around the central sphere.

---

[1] E. Lampe, *Archiv der Mathematik und Physisk,* vol. 1, 1901.
[2] Actually Hoppe considered the *centers* of the surrounding spheres, not the points of contact, but that is not important.

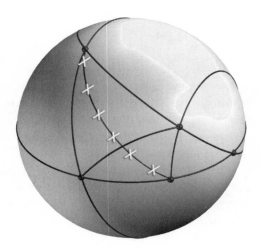

Hoppe's net

Hoppe's demonstration is tantamount to saying that thirteen balls cannot touch a central ball simultaneously. In contrast to Bender, he had made no prior assumption about the arrangement of the spheres. His demonstration was quite general.

After a year someone else took issue with both Bender's essay and Hoppe's commentary. In 1875, the paper "A stereometric problem" by Siegmund Günther appeared in the *Archiv der Mathematik und Physisk*. Günther was a geographer and geophysicist in Munich who had advanced the once-popular theory that the core of the earth consists of a gas of high density. He was also interested in molecular physics and framed the problem of the thirteen spheres as a question of how many atoms could be influenced by the forces of another atom, if these forces reach out in all directions.[3] He stated that the method employed by Hoppe was too complicated and suggested his own, simpler technique as an alternative.

Günther began his exposition by considering the shadows of the surrounding spheres that are thrown onto the surface of the central ball. Unsurprisingly, he reached the conclusion that, in principle, there is sufficient room for thirteen shadows with additional space left over. Then he went through more calculations to arrive at the verdict that his proposed simple technique is not strong enough to decide whether a thirteenth ball can or cannot fit in between the twelve that do have room. Some scientific advance, that.

That's where matters stood at the end of the nineteenth century. But

---

[3] This was written at a time when the existence of atoms was not yet verified.

then, all of a sudden, Hoppe's own contribution came under scrutiny. It turned out that his proof was incomplete.

Hoppe had suggested connecting the centers of all balls with threads. Whenever two threads crossed each other, the longer one would be deleted. His claim was that the resulting net would consist entirely of triangles. But he was wrong! We present a counterexample in the picture. First all five points are connected to each other. Then the superfluous threads are deleted. At crossing *A*, thread *m* is longer than thread *n* and must be deleted. At crossing *B*, thread *n* is longer than thread *o* and must be deleted. At crossing *C*, thread *o* is longer than thread *p* and is deleted. Finally, at crossing *D*, *q* is longer than *p* and is deleted. What is left are the five outside threads and the interior thread *p*. And that's the problem. In contradiction to Hoppe's claim, the net that remains is *not* composed entirely of triangles. It contains a quadrilateral.

The mistake invalidated Hoppe's entire proof. Further development on the question had to wait until the middle of the twentieth century, when two mathematicians finally provided the conclusive answer to the kissing problem. In 1953, Kurt Schütte from Germany and the Dutchman Baartel Leendert van der Waerden joined forces to settle the Newton–Gregory debate once and for all. The fruits of their cooperation were published as "Das Problem der dreizehn Kugeln" (The problem of the thirteen spheres) in the *Mathematische Annalen,* the foremost mathematics journal in Germany.

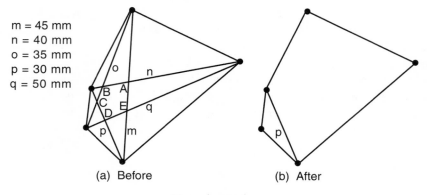

Hoppe's mistake

Schütte and van der Waerden had already met before the start of World War II at the celebrated University of Göttingen. Kurt Schütte, born in 1909, was the sixty-ninth and last Ph.D. student of David Hilbert, the most important mathematician of the first half of the twentieth century.[4] Actually, the doctoral candidate didn't see much of his thesis adviser, who was already seventy-one years old at that time. His contacts to the professor went through Hilbert's assistant, Paul Bernays, who would later become a world-renowned logician in Zürich. The twenty-four-year-old Schütte met Hilbert only once before he graduated in 1933.

After finishing his thesis, which dealt with decision problems, and working for two years as a schoolteacher, Schütte realized that the professional outlook for mathematicians was not very bright. So, during World War II, he embarked on a second career as a meteorologist. After the war Schütte taught at high schools and then slowly worked his way up through the academic ranks. The high point in his career came in 1959, when Kurt Gödel invited him to Princeton's Institute for Advanced Studies for a sabbatical. His research concentrated on issues concerning mathematical proofs and culminated in a well-regarded book, *Proof Theory,* in 1977. Schütte retired in the same year but did not let age interfere with his work. Although practically blind, he stayed active until his death in 1998. His last paper appeared posthumously in 1999.

B. L. van der Waerden was the senior of the two. He had already served as an assistant at Göttingen when Schütte was still a doctoral student. Van der Waerden was born in Amsterdam in 1903 and as a child showed prodigious talent for mathematics. After finishing high school, he went on to study mathematics at the University of Amsterdam. There was not much to teach the gifted student and he completed the required program early. With time on his hands, he decided to spend a semester at the famous University of Göttingen. It was a university of the old type. Professors were impeccably dressed, distant, and presumptuous. They delivered their lectures either flawlessly or in such a confusing manner that everyone was left mystified. Students hardly dared to address them, and lowered their eyes when spoken to by one of these demigods. It was considered an honor when a promising disciple was invited to a professor's home.

The University of Göttingen contained the world's nerve center of mathematics: the Mathematische Institut. Felix Klein, the "great Felix," as his students called him, held court at the Institut, which was also famous for

---

[4] The French would claim Henri Poincaré as the foremost mathematician of the time. At the very minimum, Hilbert was one of the *two* greatest mathematicians of the first half of the twentieth century.

being the first academic establishment to sport an open-shelf reading room. Here there were no grudging librarians who would make humble readers wait for hours on end before finally handing over a jealously guarded book.

After the first stint at Göttingen, it was time for van der Waerden's military service, and he used his spare time as a conscript in the Dutch army to write his Ph.D. thesis. He would have liked to present his work in German, so that the revered professors and colleagues at Göttingen would be able to read it, but the University of Amsterdam's rules allowed only a single foreign language: Latin. Not surprisingly, van der Waerden preferred Dutch. In 1926, the young man received a Rockefeller grant to spend a postdoctoral semester at Göttingen.

The overwhelming intellect at Göttingen was Hilbert, of course, but there were others of similar caliber, and van der Waerden met Fräulein Emmy Noether, who—to quote Albert Einstein—was "the most significant creative [female] mathematical genius thus far produced." The only woman ever to have been made *Privatdozent* at Göttingen, she revolutionized algebra and deeply influenced the young Dutchman. Based on her lectures, which were totally incomprehensible to anyone but a select few, as well as on the lectures of Emil Artin, whom he visited in Hamburg, van der Waerden wrote a pathbreaking treatise, the two-volume *Moderne Algebra,* which appeared in 1930 and 1931. The books became instant bestsellers, were immediately translated into English, Chinese, and Russian, and are still bought today by students all over the world. The two volumes made van der Waerden's name famous. As Saunders MacLane put it, "it was van der Waerden who understood the real thrust of abstract algebra and who presented it [in the two volumes] abstractly but without pedantry. . . . We are fortunate that [Noether's] imagination has been made accessible by van der Waerden."

Van der Waerden would have liked to keep the next two decades of his life under wraps. The blurbs on both volumes of his famous work only mention a post in Groningen in 1928, then gloss over the next twenty-three years, and coyly pick up again in 1951, when van der Waerden became a professor in Zürich. What happened in the period between these two positions? There is not much to be ashamed of, but there's certainly nothing to be proud of, either. Just before the start of the darkest period in recent European history, in 1931, van der Waerden was offered a professorship in Leipzig. He accepted, and the family moved east. There is nothing to show that the professor had ever been close to the Nazis. A longtime colleague of his in Zürich, Benno Eckmann from the ETH, described van der Waerden's convictions as decent but naïve. His later reluctance to speak about his personal experiences during the war years may have fostered some unpleasant questions.

The partnership between van der Waerden and Schütte took place in the early 1950s. Following their initial meetings at Göttingen, they parted ways, but after the war they linked up again to write the definitive paper on the problem of the thirteen spheres. The paper contained two proofs. Schütte had thought about the problem and his proof was prepared and ready to go when van der Waerden came up with a simpler one. What was to be done? On the one hand, it would have been a shame to let a good proof go to waste, even if it was superfluous. On the other hand, could one, in good conscience, publish a convoluted proof if a simpler one exists? Apparently not, and Schütte and van der Waerden decided to publish both proofs together.

The two gentlemen started their joint paper by making polite curtseys to each other. On the first page they emphasized that Schütte's proof, although relegated to the last section of the paper, was the first chronologically, and that this author therefore deserves priority. They continued by stressing that van der Waerden's proof, even though it saw the light of day only later, was actually the simpler one, and was therefore pulled to the front. Then they took the reader through six sections of propositions, lemmas, and theorems to prove that Newton had been right all along. Both proofs consisted of showing that a sphere that can be touched simultaneously by thirteen balls of radius equal to 1 must itself have a radius greater than 1. That was the death knell for Gregory's thirteenth ball.

By the way, how large must a ball be so that thirteen spheres *can* touch it? In an earlier paper Schütte and van der Waerden found such an arrangement with the central ball having a radius of 1.04557. Is this the smallest sphere that can be kissed by thirteen balls? It is believed that this is so, but it has yet to be proven.

Schütte and van der Waerden were not able to rest on their laurels for long. Across the Channel, the Englishman John Leech felt that the paper from the continent was too intricate, and that even van der Waerden's simpler proof was too complicated. He decided to provide an even simpler demonstration of the impossibility of thirteen spheres.

Leech was born in 1926, and educated at King's College, Cambridge, whence he graduated as a wrangler with a B.A. in 1950. ("Wrangler" is one of those esoteric blue-ribbon signs of esteem, like the "Order of the Garter," reserved for British overachievers.) Leech started his professional career by building early versions of digital computers, but returned to Cambridge as a Ph.D. student a few years later. He was one of the first mathematicians to use mainframe data processors, not just to compute numerical solutions to military or engineering problems, but to apply them to theoretical questions. He pioneered the use of computers for algebra by writing one of the early programs to implement an enumeration algorithm. For eight years he served as lecturer in the computing laboratory at Glasgow University in Scotland. In 1968 a new university was founded in Stirling, near Glasgow, and Leech was appointed head of computing science. He stayed at Stirling for twelve years, taking early retirement in 1980 because of ill health. Leech died from a heart attack in 1992 while on board a paddle steamer on a river in Scotland. On its way back to Glasgow the boat's red ensign was lowered to half-mast. Today he is best remembered for the so-called Leech lattice, about which I shall have more to say.

When Leech wrote his version of the proof for the kissing conjecture in three dimensions he didn't bother to look very far for a catchy title. Even though he set out to simplify and improve Schütte and van der Waerden's paper, he obviously liked it well enough to simply translate the title into English and use it as a heading to his own article, which then appeared as "The problem of the thirteen spheres." While this doesn't qualify as plagiarism, it does reveal a serious lack of imagination.

Leech's article, which builds on the method that Reinhold Hoppe had tried earlier, was published in 1956 by the *Mathematical Gazette*—albeit without mentioning the German mathematician's contribution. The *Gazette* is the publication of an association of British teachers and students of elementary mathematics. It may not carry a highbrow name like "archive," "annals," or "journal," but it is by no means a second-rate newsletter for mediocre schoolteachers. It publishes high-quality papers for its readership, which consists of first-rate schoolteachers, college and university lecturers, and others with an interest in the teaching of mathematics.

Leech's article was a short piece of work, only two pages. But even though it used only elementary mathematical techniques and applied them in a straightforward manner, the paper cannot be called trivial. It also can't be called an easy read. Try to read the following sentence out loud: "No two joins of this network cross, since any four points of the network form a quadrilateral of sides at least $\frac{1}{3}\pi$ whose diagonals cannot both be less than $\frac{1}{2}\pi$, the extreme case being that of the regular quadrilateral of side $\frac{1}{2}\pi$ whose diagonals are both exactly $\frac{1}{2}\pi$." It is quite amazing that such convoluted monsters could fit into two pages. Had the *Gazette*'s readership consisted of schoolteachers with an interest not only in the teaching of mathematics but also in English, the editors would have certainly added some punctuation marks here and there, and occasionally made two sentences out of one. This would not have lengthened the paper all that much, and the effort would have been well worth it.

But once Leech's paper is translated into proper English, it becomes a beauty of mathematical reasoning. Using only elementary mathematics, it arrives at a profound truth. Don't let the word "elementary" fool you, however: it must not be confused with "simple." Leech's proof employed spherical trigonometry, which is a very confusing branch of mathematics. But the proof, which is presented in the appendix, is still straightforward.

Schütte and van der Waerden had proved that in order for thirteen spheres to touch a ball, it would have to have a radius greater than one. Leech's proof shows that it is impossible to weave a net that would allow thirteen balls of equal radius to kiss a central ball. Thus it was proved in three different ways that Newton was right and Gregory was wrong: *thirteen spheres cannot touch a central sphere.* The classical kissing problem was finally solved. We now know that in one, two, and three dimensions, the Newton numbers are, respectively, 2, 6, and 12.

What about higher dimensions? I pointed out before that mathematicians have no problem working in spaces of arbitrarily high dimensions. All one has to do is use one's imagination. So how about kissing numbers in higher dimensions? Here we reach the current frontiers of mathematical knowledge. Apart from the solutions to the problem in one-, two-, and three-dimensional space, kissing numbers are only known for dimensions eight and twenty-four. For other spaces we have, at best, intervals that provide upper and lower bounds for the true kissing numbers.

As in the previous two chapters, one has to distinguish whether the balls are restricted to a lattice or whether they are free to float around in any jumbled manner. The lower bound of the interval is usually given by the best arrangement that is known for some kind of lattice. This could possibly be the true kissing number, but one won't know for certain until it has been proven. So in the meantime, all one can say about the kissing number in any higher dimension (except dimensions eight and twenty-four) is that it must be *at least as high* as the lower bound, that is, the number that has already been found for some lattice.

In 1979, two mathematicians at AT&T Bell Labs, Andrew Odlyzko and Neil Sloane, managed to take another step forward: they invented a method to compute *upper* bounds for kissing numbers. Together with the already established lower bounds, we now have intervals within which the kissing numbers of high-dimensional spaces must lie.

Odlyzko was educated at the premier engineering schools of the United States, Caltech, and MIT. His research interests cover a wide range of subjects: he has done important work in number theory, combinatorics, probability theory, analysis, cryptography, computational complexity, and coding theory. As a graduate student he spent a few summers at the Jet Propulsion Laboratory in Pasadena, and also did a stint with AT&T Bell Labs. That got him hooked, and for a time he served as head of the research department in mathematics and cryptography at AT&T Bell Labs. He is now the director of the University of Minnesota's Digital Technology Center.

Odlyzko teamed up with his colleague Sloane to investigate kissing numbers in high dimensions. N. J. A. Sloane was an old hand at balls and spheres, having written two papers with John Leech back in the early 1970s. He grew up in Australia and worked there as a student for the Postmaster General's Department. This is a fancy name for the state phone company, and Sloane had fun erecting telephone poles, splicing cables, and driving around the countryside in large trucks. When he came to the United States he was well prepared to work for another telephone company: he joined the research division of Bell Labs where he has been ever since receiving his Ph.D. in electrical engineering from Cornell. With John H. Conway from Princeton University he wrote the mathematical bestseller *Sphere Packings, Lattices and Groups,* which is considered the bible of sphere packings. *SPLAG,* as it is known to insiders, is now in its third edition, which, apart from van der Waerden's *Algebra,* is a rather rare feat for a

Andrew Odlyzko

mathematical treatise.[5] One reviewer wrote that *SPLAG* "is the best survey of the best work in one of the best fields of combinatorics, written by the best people. It will make the best reading by the best students interested in the best mathematics that is now going on." Some holier-than-thou purists claim that, like the bible, *SPLAG* contains no proofs, but Sloane doesn't agree. Proofs are provided, at least in outline, throughout the book.

But Sloane is no one-shot author, and another book has also become a hit. If you want to believe one reader of *An Encyclopedia of Integer Sequences,* then "there's the Old Testament, the New Testament, and the Encyclopedia of Integer Sequences." The Web version of the book gets eight thousand hits a day. However, Sloane's interests are not limited to sphere packings or integer sequences. In his CV he lists his membership in the American Alpine Club on the same level as the medal that was bestowed on him by the Collège de France in 1984, or the receipt of the Claude E. Shannon Award of the IEEE Information Theory Society in 1998. And a book on rock climbing in the New Jersey Crags figures in his bibliography, just below the bibles on sphere packings and integer sequences.

By the way, Conway, Sloane's coauthor on *SPLAG,* is a very important figure in the sphere packing business. He is also a real character, usually walking around with bare feet or in sandals. For thirty years he refused to go to the barber. In this he unwittingly emulates Hoppe, without suffering from the German's other shortcomings. In fact, he is usually the life of a party. He's good at card tricks and coin tricks, knows the names of all visible stars in the northern hemisphere, can tell you the correct day of the week of any date, and is prepared to recite the first thousand digits of π when given half a chance. A journalist once called him a *mathemagician.*

Conway was born in Liverpool, England, one day after Christmas in 1937. That was just a couple of years before the Beatles saw the light of day in the same town, and his father taught Paul McCartney and John Lennon chemistry at school. As a pupil Conway showed great ability in all subjects, but most of all in mathematics. After high school he went to study at Cambridge, where he got his first job as university lecturer in mathematical logic. But then his career got stuck. He was already in his late twenties and had not yet done anything to ensure his immortality. "I became very depressed. I felt that I wasn't doing real mathematics; I hadn't published, and I was feeling very guilty because of that," he wrote. He knew he was a first-class mathematician, but nobody else did. Conway was desperate to

[5] Of course, there are textbooks that make it through many printings and editions. But that's different—they are required reading.

make his mark. And then he got a break. The subject that was to provide him his entry into stardom was group theory.

Algebraic groups consist of elements, like the integer numbers −3, −2, −1, 0, 1, 2, 3, and so on, and an operation that combines any two elements, for example, addition. One requirement for a group is that whenever two members of the group are combined, like 7 and 9, the result is also a member of the group.[6] The even numbers, for example, also form a group "under addition," since adding two even numbers yields an even number. On the other hand, odd numbers do not form a group under addition, since the sum of two odd numbers, such as 11 and 17, is *not* an odd number.

The two groups mentioned (integer numbers and even numbers) have an infinite number of elements. But there are also groups that are composed of a finite number of elements. For example, the digits indicating the hours on your wristwatch form a finite group under addition. If the little hand points to 9, and you add 7 hours, the little hand will point to the number 4 on your wristwatch. Hence the requirement that the combination of 9 and 7 also be a member of the group is satisfied. The "wristwatch group" is a finite group with twelve elements.

It is one of the remarkable scientific feats of the twentieth century—almost comparable in scope to the mapping of the human genome—that the mathematical community has been able to classify *all* finite groups. It took the combined efforts of dozens of mathematicians and about five hundred separate papers with a total of more than ten thousand pages to complete the task. Victory was declared in 1982. But in the mid-1960s work on the classification of finite groups was still in its heyday. In fact, most people believed that it would take well into the twenty-first century to complete the classification. Groups had been discovered, and were still being discovered, that did not fit into any of the developed schemes. These bizarre groups were called "sporadic simple groups." (When it was all over, it turned out that there exist exactly twenty-six sporadic simple groups, and they are by no means "simple" in the usual sense.)

When Conway was trying to make a name for himself in Cambridge, John Leech had just discovered the twenty-four-dimensional grid that would henceforth carry his name. He set about investigating its characteristics. One of the important attributes of a geometric object, like a grid, is its symmetry. In the same manner that a cube can be rotated and twisted around its axes and still look like a cube, the Leech lattice can also be

---

[6] There are two more requirements: the group must contain a neutral element, like 0, and to each element, say the number 5, there must be an inverse element, like −5.

turned, revolved, and rotated—albeit in twenty-four dimensions—and still look like the original.

If a body has multiple symmetries, it can be rotated around one axis, then around a different axis, and again around the first axis in the opposite direction, and so on. Being symmetric, the body will look exactly the same after each of the rotations. This means that rotations can be added, which is precisely what is required in order for a set of elements to form a group. Hence the rotations of symmetric bodies are elements of a finite group. The characteristics a group has depend on the object itself. Leech knew that the symmetries of his lattice would be interesting, but he was aware that he lacked the group-theoretic skills necessary to investigate them. So he dangled the problem before Conway, who immediately took the bait.

Conway told his wife that this was something important and difficult, and that he was going to work on it Wednesdays from six to midnight and Saturdays from noon to midnight. He need not have planned that far ahead. It took him only a single Saturday session to crack the puzzle. What Conway discovered that evening was that the group describing the symmetries of the Leech lattice was none other than one of the sporadic simple groups that had eluded discovery until then. It contains 8,315,553,613,086,720,000 elements. Soon thereafter, the discovery of the "Conway Group" gave rise to the detection of three more then-undiscovered sporadic simple groups. That breakthrough, which took the worldwide classification effort a giant step forward, gave Conway a badly needed ego boost. He was immediately named Fellow of the Royal Society, and has been in the forefront of mathematics ever since. In 1986 he joined the faculty at Princeton University.

Incidentally, the Conway Group, large as it may seem, is by no means the largest sporadic group. The appropriately named Monster Group, which was discovered by Robert Greiss in 1980, has 808,017,424,794,512,875,886,459,904,961,710,757,005,754,368,000,000,000 elements. It contains more elements than there are particles in the universe. The Baby Monster, which has only 4,154,781,481,226,426,191,177,580,544,000,000 elements, is still several sizes larger than the already enormous Conway Group. Mathematicians, who are not easily disconcerted by bizarre objects, consider sporadic simple groups very weird indeed.

With the discovery of the sporadic group, Conway had made his name and from then on he could do whatever he liked. And he did. One of his preoccupations was the ancient Chinese board game Go. That interest led in some roundabout way to the discovery of surreal numbers. Then he invented the Game of Life, a computer simulation of cellular automata. Its rules are extremely simple but the evolving game gives rise to amazingly complex behavior. According to author Martin Gardner, people who got hooked on the game "consumed millions of dollars in illicit computer time" to play it while at work.

Apart from his work on group theory, Conway has also made advances in number theory, game theory, coding theory, tiling, knot theory, and the theory of quadratic forms. He does not only write esoteric works like *SPLAG*. His popular books *On Numbers and Games, The Book of Numbers, Winning Ways for Your Mathematical Plays, On Being a Department Head,* and *The Sensual (Quadratic) Form* (some coauthored with colleagues) have consistently scored positive reviews. Of course, he also knows a thing or two about geometry. Tom Hales, who will accompany us in this book's final chapters, heard about Kepler's conjecture for the first time as a graduate student in Cambridge, when Conway mentioned the unsolved problem in one of his lectures.

The method that Odlyzko and Sloane invented to compute the upper bounds for kissing numbers relies on polynomials that fulfill certain requirements. They computed the upper bound in four dimensions to be 25. On the other hand, a lattice arrangement is known (the so-called *laminated* lattice) in which 24 balls kiss the central sphere. Hence, a modern version of the Newton–Gregory debate can be formulated as follows: Is the maximal number of white balls that can kiss a black ball in the center of four-dimensional space 24 or 25? In five dimensions the Newton number lies somewhere between 40 and 46, in dimension six it is at least 72 and at most 82, and in seven dimensions the interval is bounded by 126 and 140 balls.

Then, in eight dimensions, all of a sudden the Newton number is known exactly: 240 white balls can kiss the black ball in the center. Why is that number known precisely? In this case the upper bound came out to be 240, but a certain, well-known lattice arrangement, called E_8, also allows 240 balls to touch the central sphere. Since the actual kissing number of E_8 coincides with the theoretical upper bound, 240 must be *the* highest kissing number. From dimensions nine to twenty-three, again, only bounds are known.

An interesting case occurs in dimension nine. Up to this point, everything we know, or assume to be correct, happened on lattices. For example, the densest arrangement of disks on the plane occurs when coins, say, are placed in a regular hexagonal pattern. The same holds for circle coverings. Kepler conjectured that the densest packing of spheres occurs when the apples or oranges are stacked in a regular market stall arrangement. Even the highest kissing number in three-dimensional space occurs on a regular grid,[7]

---

[7] As we saw in chapter 5, however, "12 around 1" can also occur on icosahedra, which cannot be combined into grids.

and as we just saw, the same is true for the eight-dimensional kissing number. But suddenly, with dimension nine, something unexpected occurs: the highest possible kissing number for lattices is 272, but a jumbled-up, nonlattice packing exists in which a sphere can touch 306 others! So for the first time we encounter a nonlattice arrangement that is superior to any possible lattice arrangement.[8] This example shows that nothing—repeat, nothing—may ever be taken for granted in mathematics. Everything may go along nicely and smoothly until in dimension nine the exact opposite of what we expect should have happened happens.

In dimension twenty-four we once again have an exact result for lattice packings: 196,560 white balls can kiss the twenty-four-dimensional black ball in the center. It was the lattice that John Leech discovered in 1965. Hence, at least that many balls can kiss a central sphere in twenty-four-dimensional space. Fourteen years later, Odlyzko and Sloane's method showed that at most that many balls can kiss a central sphere. When the upper bound coincides with the lower bound—kaboom!—we have found the Holy Grail, the highest kissing number in twenty-four dimensions. (The mathematician V. I. Levenshtein of the Russian Academy of Sciences discovered the kissing number for twenty-four dimensions at the same time as, but independently of, Odlyzko and Sloane. His feat was all the more remarkable since he had no computers available to him.)

The discovery and description of certain grids in high-dimensional space are Leech's most lasting achievement. His pursuit of high-dimensional excitement started with a paper in 1964 in which he discussed sphere packings in eight or more dimensions. A year later, a supplement followed about a twenty-four-dimensional grid that would henceforth carry his name: the Leech lattice. It allowed a sphere packing that was locally twice as dense as the original. Since then it has been shown that, in twenty-four dimensions, the Leech lattice arranges balls locally in the absolutely densest manner (that is, there is neither a lattice nor a nonlattice arrangement that locally packs balls closer). Furthermore, it has been conjectured that the Leech lattice may also globally represent the densest packing in twenty-four dimensions. In this book we will not delve further into this area; we'll just stick to Kepler's three dimensions.

By the way, in 128-dimensional space there exists a grid that allows 218 billion balls to kiss *one* ball in the center. Quite a crowd, you may say. But if one doesn't care much about neatness, there is a nonlattice arrangement that allows at least 8,863,556,495,104 balls (that's eight trillion and change)

---

[8] Whether 306 is also the absolutely highest kissing number is not yet known.

to touch the center ball. Neither of these numbers is thought to be the last word on the subject, however.

The debate that Newton and Gregory started in 1694 is still going strong. Consider this: A preprint of an article appeared in 1996 in which the authors established an upper bound on the degree of the polynomial that must be used to compute an upper bound for the kissing numbers in various dimensions. The paper does not reveal kissing numbers—that would be too much to ask for—but just gives an indication about the polynomial that could, in turn, be used to compute not the kissing number itself but just an upper bound for it. This is how intricate mathematics can be; every minute hint is of interest. Sherlock Holmes would have felt dumbfounded with such farfetched clues.

CHAPTER 7

Twisted Boxes

In 1831 a book by an obscure professor of physics and mathematics appeared that is connected in a roundabout way to our problem. The book itself did not have much of an impact on mathematics, but a review of it did. Ludwig August Seeber (1793–1855) left no great mark on either physics or mathematics. Just three pieces of his work are known. One of them dealt with the structure of solids and is still sometimes, although rarely, mentioned in the contemporary literature in crystallography. Another one is totally forgotten. His third work, a book he published in 1831, was called *Untersuchungen über die Eigenschaften der positiven ternären quadratischen Formen* (Investigations into the properties of positive ternary quadratic forms). It is Seeber's main contribution to mathematics, and it is this work that concerns us in this chapter.

Seeber's long-winded investigation into quadratic forms was very commendable, but extremely tedious. Nevertheless, it caught the eye of the greatest mathematician of his time, the giant of the University of Göttingen, Carl Friedrich Gauss.

Gauß (his name in proper German), was born on April 30, 1777, in the German city of Braunschweig. His father, Gebhard Dietrich Gauss, was a domineering man—"authoritarian, uncouth and unrefined" in Carl Friedrich's words—who never gained the trust of his son. His mother, however, an intelligent but semiliterate woman—she could read but not write—would be her son's devoted supporter throughout her life. The senior Gauss tried his hands at various jobs in order to make ends meet. At some time or another he earned his living as a mason, a butcher, a gardener, and a canal worker. The family lived in modest circumstances. The only relative with any even modest intellectual gifts was the mother's brother, a master weaver.

In elementary school Carl Friedrich showed extraordinary ability. One day, when he was eight years old, the teacher tried to keep the kids busy for a while and told them to add all the numbers from one to one hundred; in

C. F. Gauss

the meantime he would enjoy a cup of tea. The precocious Gauss immediately noticed that 1 plus 100 equals 101, that 2 plus 99 equals 101, and so on until 50 plus 51, which also equals 101. So there are 50 pairs of numbers whose sums always equals 101. The exasperated teacher, who had not even been able to take a sip, must have been quite annoyed when the little squirt walked to his table with the correct result, 5050, after barely a minute.

It was much to the credit of the poor teacher, J. G. Büttner, and especially his assistant Martin Bartels, that Gauss's gifts were spotted. It was no easy feat to single out the one boy, from among the fifty or so unruly kids who sat in a class, who would advance science by leaps and bounds. Bartels, later professor of mathematics at the University of Kazan, gave him special instruction (as well as he could), provided him with books, and brought him to the attention of the authorities.

When Gauss reached the age of eleven, he entered the *Gymnasium,* the high school, contrary to his father's wishes that he start earning a living. He excelled at mathematics and languages, and was promoted to the top class within two years. When he was fourteen, he was presented at court to Duke Carl Wilhelm Ferdinand von Braunschweig-Wolfenbüttel. The duke was so impressed with this kid that on the spot he awarded him a scholarship of 10 *thalers* per annum. The award would regularly be renewed and increased for the next sixteen years and kept Gauss free from financial worries until he was thirty years old. Out of gratitude toward the duke, Gauss became a lifelong supporter of the monarchy.

At age fifteen, Gauss entered the college in Braunschweig, the Collegium Carolinum, where he read Newton's *Principia* and all the works of Euler and Lagrange that he could find at the library. But the institute's book collection was sadly lacking and the young student started to look elsewhere for a more complete library. Three years after he had entered the college, and still without a diploma, he transferred to the University of Göttingen.

The duke would have preferred Gauss stay in Braunschweig, but relented after he was told that the Göttingen library had a quarter of a million books, and that its modern catalogues and liberal lending policy made it the foremost research library in Europe. He even continued paying him the stipend. Gauss devoured the books available at Göttingen. He also used the time to acquaint himself intimately with individual numbers and spent long years of endless calculations, aimless manipulations of numbers, and computation of puzzling tables.

As good as the library was, the professor of mathematics, Abraham Gotthelf Kästner, was not very inspiring. He was of mediocre talent and Gauss quickly became disenchanted. After a while, he no longer bothered to go to Kästner's lectures, as they were too elementary. The linguist, on the other hand, a man by the name of Heyne, was more impressive. The young student had to make a difficult career decision: math or languages?

While the nineteen-year-old was still battling with this question, he made a momentous discovery. He showed that it was possible to construct, with ruler and compass, a heptadecagon, which is a fancy word for the regular, seventeen-sided polygon. He also showed how it could be done.

For two thousand years it was not known which polygons could be so constructed. It was known how to construct a regular triangle, square, and pentagon. And since it is easy to divide an angle by two, the number of sides could be doubled and redoubled. Hence the 6-gon, 8-gon, and 10-gon (hexagon, octagon, and decagon), and the 12-gon, 16-gon, and 20-gon could also be constructed. But what about the 7-, 9-, 11-, 13-, 14-,

15-, 17-, and other *gons?* Can they or can they not be constructed with a compass and circle? The problem of constructing a regular *n*-gon is equivalent to dividing 2π—that is, the circle's 360 degrees—into *n* equal parts. That's no problem on a pocket calculator, which gives six or eight digits after the decimal. Surely that is more than enough for general usage, but it was not at all sufficient for the nitpicking Gauss. He wanted an exact result. Furthermore, Gauss did not even have access to a slide rule.[1] And measuring 21²¹⁄₁₇ degrees with the help of a protractor would have been cheating. Only a ruler and a compass were allowed.

The solution came to Gauss while still in bed one morning, just after he woke up: a regular polygon may be constructed by ruler and compass if the number of sides, *n*, is equal to $2^k pqrs \ldots$, where *k* can be any integer, and *p, q, r, s, . . .* are prime Fermat numbers. No other regular polygons can be constructed by ruler and compass.

Fermat numbers are numbers of the form $2^{2^m} + 1$. Only five Fermat numbers are known that are prime: 3, 5, 17, 257, and 65537, corresponding to *m* = 0, 1, 2, 3, and 4. For *m* = 5, the Fermat number is 4,294,967,297, which isn't a prime number: it can be factored into 641 and 6,700,417. I won't even write down the Fermat number for *m* = 6 since it has thirty-nine digits. But believe me, it's no prime number either, because it factors into 59,649,589,127,497,217 and 5,704,689,200,685,129,054,721.

These five Fermat numbers, together with the number 4, are the basis for Gauss's early-morning musings. What he found was that any polygon can be constructed whose number of sides is one of the five known Fermat primes (and the number 4), or a multiple by 2^m, or a combination of these numbers. (But each number must appear only once.) For example, the 12-gon is constructible as four times the triangle, the 15-gon is a combination of the triangle and the pentagon, and so on. Thus it is possible to construct by ruler and compass any regular *n*-gon if *n* equals 3, 4, 5, 6, 8, 10, 12, 15, 16, 17, 20, 24, 32, ... (It is *not* possible to construct the 18-gon, because in the product $2 \cdot 3 \cdot 3$, the number 3 appears twice.)

When Gauss showed his 17-gon to Abraham Kästner, the professor didn't even understand what Gauss was talking about. But *Hofrat* (councilor) E. A. W. von Zimmermann, a professor at the Collegium Carolinum,

---

[1] Slide rules were invented toward the middle of the seventeenth century. There is no record, however, of Gauss having used them.

was extremely pleased with the student's finding and announced it in the *Intelligenzblatt der allgemeinen Litteraturzeitung* (Intelligence report of the General Literature Gazette) under the heading *"Neue Entdeckungen"* ("New Discoveries"). Gauss's achievement was to reduce the geometrical problem to a question in number theory. By the way, Gauss also stated with emphasis that no other polygon could be constructed, but he provided no proof. That question was only settled forty years later, by P. Wantzel, who proved that it is impossible to construct polygons with 7, 9, 11, 13, 14, 18, 19, 21, 22, 23, 25, 26, 27, 28, 29, 30, 31, . . . sides.[2]

Apart from settling the question of which polygons can be constructed geometrically, this feat also settled Gauss's career decision: languages were out, mathematics was in. One may but wonder what great achievements Gauss would have accomplished in the study of the Latin language, or what Greek aphorisms are forever lost because he abandoned the subject. But the philologists' loss was, is, and will remain the mathematicians' gain.

After his success with the heptadecagon, the ideas came so fast that he did not even have time to write them down. His diary is full of mathematical discoveries that he never bothered to bring to fruition. Many of the half-baked ideas that he scribbled into his notebooks would have sufficed other men as the work of a lifetime.

In 1799 Gauss received his doctoral degree in absentia, even before the publication of his thesis, in which he gave a proof of the fundamental theorem of algebra. The fundamental theorem states that any polynomial with complex coefficients has a root in the field of complex numbers. So great was Gauss's stature already that he wasn't required to give the customary oral defense of his thesis. Two years later, when he was only twenty-four years old, Gauss published *Disquisitiones Arithmeticae,* a book that dealt with what we today call number theory. Its content was not quite understood by his contemporaries, but it immediately catapulted Gauss to prominence. One of the sections dealt with quadratic forms that Lagrange had discovered earlier (see chapter 3), but whose importance the Italian-Frenchman had failed to appreciate.

In the same year Gauss accomplished another feat. An astronomer by the name of Giuseppe Piazzi had discovered an asteroid in the sky in the preceding winter. He named it Ceres, and then promptly lost sight of it. Piazzi and others wiggled their telescopes in all directions of the firmament, in the hope of finding it again. But it was in vain; the asteroid was lost. Along came Gauss, who managed to calculate the orbit of the tiny and distant asteroid

---

[2] Wantzel also proved that no angle can be perfectly trisected, except if it is $90° \cdot m/2^n$.

based on the the few observations that had been recorded by Piazzi.[3] The astronomers stopped wiggling and focused their telescopes to the point that Gauss had predicted. That point was way off from where they thought Ceres should have been, but they promptly rediscovered it. Now Gauss's fame as an astronomer was also established. The young man had used a secret weapon for his calculations. He had developed the method of squared errors to determine the asteroid's orbit in spite of the paucity of observations. But, like Newton, who had managed to decipher planetary motion without revealing calculus, Gauss kept the method of smallest squares secret.

In 1806 Gauss's benefactor, the seventy-year-old duke of Braunschweig-Wolfenbüttel, was called up to fight Napoleon. Against his will, the Prussian cabinet had named him supreme commander of the army. But the duke, who forty years earlier had been a successful general in the service of King Frederick the Great, Leonhard Euler's erstwhile boss, was too old and frail to take command of a disorganized army that lacked a clear chain of command. The Prussians were badly beaten in the battles at Jena and Auerstaedt. The duke suffered mortal wounds and died a few days later.

Gauss was distraught. To him the duke had been one of the noblest representatives of an enlightened monarchy, while Napoleon symbolized the worst dangers of revolution. Gauss's conservative tendencies were reinforced. Henceforth he refused to write in French, and feigned ignorance of the language when circumstances forced him to interact with Frenchmen. Actually, he didn't just dislike French mathematicians, his antipathy extended toward fellow mathematicians in general. He treated many contemporary colleagues as if they came from another world and lamented the "shallowness of contemporary mathematics."

With the duke gone, Gauss had to find other ways to assure his financial security. He did not cherish the thought of teaching and preferred a job where he could concentrate on his research. Fortunately, the Ceres feat had made his name known to the authorities. Gauss decided on a career change. A year after the duke's demise, Gauss landed a job as the director of Göttingen University's observatory. From then on he was a professional astronomer. He kept the position at the observatory until his death, half a century later.

Since his times with the revered Duke von Braunschweig-Wolfenbüttel, Gauss remained a deeply conservative and nationalistic German patriot. Even though he read many books in various languages, his cultural outlook remained narrow. And he usually did not like to take a stand. When King

---

[3] In 1978 radio signals reflected by Ceres were detected again by astronomers.

William IV of England and Hannover died, Queen Victoria followed him to the throne in England, but the chauvinist laws of Hannover did not allow a female ruler. Therefore, Victoria's uncle, the duke of Cumberland, became king of Hannover. One of his first decrees was an annulment of the oath to the constitution, sworn by the professors of the University of Göttingen. This raised the anger of the faculty, and seven prominent professors wrote a manifesto to protest this edict. Among the Göttingen Seven (G-7) were Gauss's friend and colleague Wilhelm Weber, and the orientalist Heinrich Ewald, husband of his oldest daughter, Minna. The king was not perturbed by the protest since professors were not difficult to come by. They could be hired as easily as ballerinas, he remarked, and had the G-7 kicked out of Göttingen. For Gauss, the loss of Weber, who was not just a colleague but also like the son he always wished for, was irreparable, and the move of his favorite daughter away from Göttingen left a deep void. But while the controversy was at its height, he never lifted a finger to help his tormented colleagues. Maybe his conservative tendencies forbade any involvement in matters that concerned the royal authorities, or maybe he was just reluctant to meddle in politics. Whatever the reason, Gauss's timid status as a bystander and his disloyalty toward his colleagues, does not compare well with, say, Isaac Newton's strong stand against the Crown's interference in the matters of the University of Cambridge.

After Seeber's book came out, it was only good form that Gauss, the author's former teacher, should write a review of it for the *Göttingische gelehrte Anzeigen* (Learned announcements of Göttingen). The *gelehrte Anzeigen* was, and still is, a review journal in which scholars in philology, philosophy, theology, science, mathematics, and other disciplines review the books of their rivals. Founded in 1739, the *Göttingische gelehrte Anzeigen* was published, in Gothic script, under the auspices of the Prussian Royal Society of Science. During the eighteenth and nineteenth centuries it was the foremost review journal of the world. The leading savants of Germany used the *Anzeigen*'s pages to heap praise on colleagues, or to tear their work apart. The journal was considered an emblem of distinction for the University of Göttingen. Today the *Anzeigen* is put out by the Academy of Science of Göttingen, and is the only review journal in existence with such a long history. The reviews are serious affairs, and give the professors the opportunity to critically discuss and put into context the ideas presented in their colleagues' works. Very often one needs to be a specialist just to read them.

Gauss was an active reviewer. Even though he was a strict and often unfair critic of the work of others, his book reports were usually mild and he was often inclined to dispense benevolent praise. Except, that is, for remarks to the effect that he had known everything all along.

Gauss's review of Seeber's work appeared on July 9, 1831, only a few months after the book was published. Gauss didn't sign the review, since this was not customary at that time, and articles were acknowledged, at most, with the author's initials.[4] The "publish or perish" syndrome of modern-day universities was not yet the norm and it didn't matter how many articles a professor published. Fame and recognition came with truly original work that managed to stand the test of time. Gauss did not even bother to put his initials to the review, but insiders were aware who had written it, especially since the *Disquisitiones Arithmeticae* was frequently mentioned. And everyone knew who the author of *that* piece of work was.

The review was quite laudatory, if a little condescending. At one point Gauss stated that he had actually done the lion's share of the work in his own *Disquisitiones Arithmeticae,* thirty years earlier, but that Seeber's tedious efforts were praiseworthy nevertheless. Further on, he wrote that it would take only a few words to summarize what was new in Seeber's book. Some compliment. According to Gauss, Seeber's primary contribution was the tying together of some loose ends that he himself had left dangling in *Disquisitiones Arithmeticae*—but only because he had not been interested in those details at the time.

The main thrust of Seeber's work is the description of a method that allows the transformation of any quadratic form into its reduced state. This is followed by the proof of a theorem that states that there exists exactly one reduced quadratic form in each class of equivalent forms. Gauss considered quadratic forms one of the most interesting and richest subjects of higher arithmetic, that is, number theory, because of the many applications outside of its field. He mentions crystalline structure as one example where quadratic forms come in handy.

The review started with a warning to potential readers: the description of the reduction method covers 41 of the 248 pages of the book and the proof of the theorem takes another 91. This inordinate length may turn readers off, Gauss wrote, only to add immediately that his remark should by no means be understood as a reproach. To the contrary, he emphasized, the first person who shows that a proof exists and that a statement is true must always be thanked, no matter how many pages it takes. Nobody should complain about a long proof unless he himself provides a shorter or more elegant one.

Before he started on his ninety-one-page proof, Seeber wanted to get an upper bound for the parameters of the ternary quadratic form that he

---

[4] This tradition is nowadays kept alive by the Swiss newspaper *Neue Zürcher Zeitung.* Yours truly signs his reports in the *NZZ* only as "gsz."

would have to manipulate. He was especially interested in a certain ratio, whose importance for sphere packings will become apparent later on. (The ratio consisted of the square of the product of the ternary quadratic form's first three parameters, *abc,* divided by the quadratic form's determinant.) To get a feel for the size of this ratio, Seeber started out by calculating many examples. He never came across a single instance where the ratio was greater than 2. After six hundred examples the suspicion befell the perceptive professor that maybe the ratio is *always* smaller than 2. But even if he had computed a thousand, ten thousand, or a million ratios, and not found a single one that contradicted his suspicion, he could still not have stated this fact as an absolute truth. He had to prove it. Unfortunately, Seeber was unable to do so. He tried, all right, but managed to prove only that the ratio had to be smaller than 3. Finally the frustrated professor gave up. He published his book with the proof of the weaker theorem, recorded his suspicion that the ratio is actually always smaller than 2, and let it go at that.

This is where Gauss picked up the cue. Referring to Seeber's conjecture very politely—and with only the slightest trace of contempt—as a "peculiar theorem found by way of induction," he started taking an active interest in it. He did not refer to "induction" in the word's modern, mathematical sense. Rather he meant that Seeber had ventured a lucky guess made after six hundred tests. But now the prince was going to "make a contribution in this review to the completion of the theory." Using only the simplest of arithmetic manipulations, and requiring not more than a page and a half, Gauss furnished a very clever proof of the fact that the ratio under discussion is always less than 2.

To prove Seeber's conjecture, Gauss wrote down an equation whose right-hand side consisted of a few simple terms made up of the quadratic form's parameters. Then he showed that each one of the right-hand terms is positive. This, of course, implies that the left-hand side is also positive, which—as the appendix to this chapter shows—meant that the ratio is always smaller than 2. And, voilà, QED.

Gauss's review of his book must have felt to Seeber like a slap in the face. He had computed more than six hundred ratios, written a 248-page book, proved a weaker theorem, but never managed to achieve the aim he had set for himself. And here, in just forty lines, Gauss provided the elusive proof that had frustrated Seeber for so long. Granted, a slap in the face by the *princeps mathematicorum* may not sting as much as a slap by any other mortal, but a slap it was.

But the best is yet to come. Gauss's strange numerical result about some esoteric ratio has an extremely important implication for the sphere packing problem. I mentioned before (and also in chapter 3) that the determinant of a binary quadratic form equals the surface of a grid's fundamental cell. Joseph-Louis Lagrange, who pioneered the study of quadratic forms, had not realized that. He thought that he simply dealt with higher *arithmetic*. But Gauss saw straight through the number-theoretic problem and succeeded in peeking further afield, toward geometry. He pointed out, in the review of Seeber's book, that quadratic forms have a *geometric* interpretation: the square root of a quadratic form's determinant is the volume of a lattice's fundamental cell. This interpretation holds both in the plane (for binary quadratic forms), and in space (for ternary quadratic forms).

So how is the theorem that was conjectured by Seeber and proved by Gauss, related to the packing problem? Let us look at a three-dimensional, right-angled lattice whose edges are a, b, and c. Obviously, abc is the volume of an upright box. Change the lattice by turning and twisting the axes, thereby squashing the box and reducing its volume. Now comes the interesting part: the ratio that had caught Seeber's attention was $a^2b^2c^2$, divided by the determinant of the lattice's quadratic form (i.e., the square of "volume of the upright box" divided by the square of "volume of the twisted box"). Actually, it was Gauss who gave the two expressions their geometric interpretations. And then he showed that the ratio is always smaller than 2. What does that mean? Taking square roots on both sides of the equation, Gauss's theorem states that the volume of the twisted box can never be reduced by more than 29.3 percent from the volume of the upright box, no matter how the axes are twisted, turned, and squashed (see the appendix).

We now make another important observation. Every fundamental cell contains parts of eight spheres—one at each corner. When these parts are joined together they always add up to one complete sphere, no matter what the cell's shape. (Recall the pizzas in chapter 3.) The smaller the volume of a fundamental cell, the denser the packing. The only requirement is that the edges be sufficiently long so that the spheres do not overlap.[5] How long is sufficiently long? In order for two spheres of radius 1 to have sufficient room at the corners, all edges must at least be length 2. So the volume of the upright box is $2 \cdot 2 \cdot 2 = 8$. On the other hand, Gauss showed that the volume of a twisted box cannot be reduced by more than 29.3 percent. Therefore, the volume of the smallest box cannot be less than 5.657. Hence, the density of a lattice packing, which is calculated as the volume of the complete sphere, 4.189,[6] divided by the volume of the smallest box, can never be greater than $^{4.189}\!/_{5.657} = 74.05$ percent.

 ❖ ❖ ❖

[5] The angles must also sufficiently large, but I won't go into detail here.
[6] $\frac{4}{3}\pi = 4.189$.

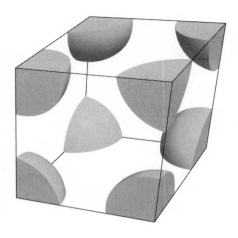

Fundamental cell in three-dimensional grid system containing the eight parts of spheres

So the ratio in question is always smaller than 2, which proves that the density of a lattice packing can never be greater than 74.05 percent. Does that number seem familiar? If it doesn't, look again at the appendix to chapter 2: 74.05 percent is exactly the density that was conjectured by Kepler to be the tightest sphere packing possible. So Gauss's observations in the review of Seeber's book prove that no three-dimensional lattice sphere packing can be denser than 74.05 percent. Kepler was right—at least for lattice packings. And which lattice arrangement provides this density? The arrangement suggested by Kepler, the face-centered cubic packing (FCC), does. But is this the *only* lattice packing of spheres that provides this density? The answer is yes, as is also shown in the appendix.

How did Gauss hit upon just the terms that would prove his point? After all, there exist innumerable possible combinations of parameters, but he picked just the ones that would work. We can be certain that Seeber had also tried many combinations, without success. Gauss simply had phenomenal insight. Fourteen years earlier he had written that "it is characteristic of higher arithmetic that many of its most beautiful theorems can be discovered by induction with the greatest of ease but have proofs that lie anywhere but near at hand, and are often found only after many fruitless investigations with the aid of deep analysis combined with luck." So it was intense meditation and a bit of luck that did it. And Gauss's penchant to get to know numbers individually probably helped. But it still seemed like a mystery. In 1850, the French mathematician Charles Hérmite referred to

Gauss's equation as "this astonishing transformation" and noted that it "seems to be based on hidden principles that I have not been able to find, in spite of my greatest efforts." His colleague V. A. Lebesgue did not agree: "I don't believe that [Gauss's] proof rests on hidden principles," he wrote in 1856, but then goes on to acknowledge that, "Gauss's proof is still the simplest one." Fascinated by the subject, Charles Hérmite returned to it in 1874 to give another proof, and then admits that it "is far removed from the elegance and depth of Gauss's demonstration." Throughout the nineteenth century mathematicians tried to improve on Gauss. Apart from Hérmite and Lebesgue, Gustave Lejeune Dirichlet gave an alternative proof (1850), and so did A. N. Korkine and Egor Ivanovich Zolotareff (1873), E. Selling (1874), and Hermann Minkowski (1883). Interest continued into this century, quite in the spirit of Gauss, who thought that "the finding of new proofs for known truths is often at least as important as the discovery itself." In 1940, Kurt Mahler re-proved Seeber's theorem, and in 1992, 160 years after Gauss's book review, yet another article appeared with the title "On Gauss's Proof of Seeber's Theorem."

As mentioned, Gauss also pointed out the close connection of quadratic forms to the structure of crystals. After all, the atoms of crystals lie on lattices. All of this Gauss only mentioned toward the end of the review, as an aside. He simply wanted to emphasize the importance of quadratic forms for disciplines other than number theory. Gauss's review was deemed so important that it was reprinted nine years later, in 1840, in *Crelle's Journal*. August Leopold Crelle was an engineer in Berlin with a soft spot for mathematics and mathematicians.[7] In 1826 he founded a journal, named it after himself, and edited it until his death in 1855. The journal, which appears monthly, continues publication to this day. High-quality articles are published in all fields of pure and applied mathematics. As a tribute to the changing times, it is now named more appropriately *Journal der reinen und angewandten Mathematik* (Journal of pure and applied mathematics). According to the publisher, it is the oldest mathematics periodical still in existence and boasts one of the largest circulations worldwide. It's not the cheap price that encourages such wide distribution. A subscription costs $2,295 per year. You can buy single issues if you don't want to commit to a full year, but it's not the kind of periodical you'd get at your local newsstand. For an individual copy you must shell out $238. At 240 pages an issue, that's just about a dollar a page.

---

[7] He was an avid supporter of the mathematician Niels Henrik Abel, who died in 1829, not yet twenty-seven years of age.

Only after the appearance of Gauss's book review was it realized that Lagrange had already given the ingredients for the solution of the two-dimensional lattice packing problem (see chapter 3). Nearly sixty years earlier, Lagrange had manipulated binary quadratic forms without realizing their connection to lattices. It was Gauss who found the missing link: binary quadratic forms are associated with two-dimensional lattices, ternary quadratic forms with three-dimensional lattices. So with one fell swoop the two- and three-dimensional lattice packing problems were solved. There is one more thing to note: the famous book review again showed how different mathematical disciplines are interwoven. Using purely number-theoretic tools, Gauss proved a geometrical theorem.

With this, the first part of Kepler's problem had been solved. Gauss showed that no lattice arrangement exists that could provide a denser arrangement than the one Kepler had conjectured. But what about jumbled-up arrangements? Even after Gauss's achievement, this question still remained unanswered, and the coming chapters will deal with the really tough question, the general, nonlattice packing problem in three dimensions.

CHAPTER 8

No Dancing at This Congress

A fter the review of Seeber's book by Gauss, Kepler's conjecture went into hibernation for close to seventy years. Gauss had shown that if the balls had to lie on a lattice, no arrangements existed which could be denser than the ones proposed by Kepler. And that's where matters stood. The question of whether a jumbled-up arrangement could be even denser was only revived at a congress of mathematicians in 1900. On August 8, the German mathematician David Hilbert addressed the Second International Congress of Mathematicians, held in Paris. He thought that open problems were the sign of a subject's vitality, and under the title "Mathematical Problems" presented twenty-three unsolved problems whose solutions, he believed, would give impetus to important new discoveries during the new century. One of the problems, number 18, dealt with Kepler's conjecture.

At the turn of the century, Hilbert was considered the most important mathematician of his time, with the possible exception of the Frenchman Henri Poincaré.[1] Hilbert was the elder colleague and close friend of Hermann Minkowski, who, through the development of the geometry of numbers, provided lower bounds for the packing density in any dimension (see chapter 4). He was also, respectively, Ph.D. adviser and host to the two-man team who proved that the highest kissing number in three dimensions is 12: Karl Schütte and B. L. van der Waerden. And he taught at the same prestigious university as Carl Friedrich Gauss, the University of Göttingen. So Hilbert definitely had some affinity to sphere packings.

Hilbert was born in 1862 in the German city of Königsberg (today Kaliningrad). His talents were recognized early, but at the University of Königsberg it was not he who outshone everyone else, but his younger col-

---

[1] A translation of his name would be "squared point." What a misnomer for a mathematician.

league Hermann Minkowski.[2] After Minkowski won the Paris Academy
prize, Hilbert's father, a Prussian judge, told his son that to presume friendship with such a genius bordered on impertinence.

But Hilbert forged ahead. After receiving his Ph.D. from the University
of Königsberg, he was called to the chair of mathematics at the University
of Göttingen in 1895. Two years earlier, the German Mathematical Society
requested a report on the state of the theory of numbers, and he wrote his
first major treatise. The *"Zahlbericht"* (translated into English as "The theory of algebraic number fields") was a brilliant synthesis of all prior investigations in number theory. More important, it also contained new concepts
and ideas that would influence research efforts for years to come. Then he
put geometry on a formal axiomatic setting. His book, *Grundlagen der
Geometrie* (Foundations of geometry) appeared in ten editions and had the
greatest influence on the subject since Euclid's *Elements*. But that was not
all. Hilbert dealt with mathematical logic, invariant theory, functional
analysis, integral equations, the calculus of variations, and even mathematical physics. In each of these subjects he provided penetrating insights. By
and by, he became the undisputed master of German mathematics.

For a while it was even believed that he, and not Albert Einstein, might
have been the father of general relativity theory. Einstein had submitted the
final revision of his article to the Preussische Akademie der Wissenschaften
(Prussian Academy of Sciences) on November 25. Hilbert, on the other
hand, had submitted an article to the Gesellschaft der Wissenschaften (Society of Sciences) in Göttingen on the same subject five days earlier, on
November 20, 1915. However, as we might expect of the orderly Prussians,
Einstein's article appeared almost immediately, on December 2. The Göttingen society was slower and Hilbert's paper saw the light of day only on
March 31 of the following year. But priority is based on the date of sub-
mission and not the date of publication, and so the prize would seem to
belong to Hilbert. Worse still, it turns out that Einstein was in possession of
Hilbert's paper before he finished revising his own article. He admitted as
much in a letter to Hilbert, on November 18, in which he thanks his colleague for sending him a copy of the paper. So much for the facts.

Did Einstein make use of material that he gleaned from Hilbert's article
during the week between November 18 and November 25? Did he plagia-
rize Hilbert's work? Recent scrutiny of the archives in Göttingen shows
that actually it may have been the other way around. Among the papers
found were the galley proofs of Hilbert's paper. They were stamped
December 6, and show that two crucial terms of the equations, which were

---

[2] See chapter 4.

present in Einstein's paper, were missing from Hilbert's proof sheets. However, in the final publication of March 1916, the two terms were present. So the uncomfortable conclusion must be that Hilbert added corrections to the galleys after he read Einstein's paper of December 2. This is quite acceptable, but the confusion could have been avoided had Hilbert done what is customary in such cases: add a byline with the date of the final revision. This would have avoided the unnecessary priority dispute, which was, it must be noted, only carried out by later historians of science and never by the protagonists themselves.

The glory of the Mathematische Institut at Göttingen came to an end in the 1930s, when the Nazis brought the faculty down. Jewish professors and assistants, who made up a large part of the mathematics faculty at Göttingen, were forced out of the university and had to flee abroad. The final end came when Hilbert's eminent colleague Edmund Landau, who had refused to read the writing on the wall, wanted to give an introductory calculus course. Until then, he had been permitted to give advanced courses, but the Nazis did not allow beginners to be taught by a Jew. Hooligans in brown shirts called a boycott and refused him entry to the auditorium.

Hilbert, who by then was in his seventies, stayed on as a professor emeritus. But the place would never be the same again. At one time, Hilbert had to provide proof that his first name, David, was no indication that he was of Jewish descent or that he was anything less than an Aryan. At a banquet, a Nazi minister asked him whether mathematics at Göttingen had suffered after it had been purged of the Jewish influence. The minister expected to hear that to the contrary, the subject had started to flourish once the devious and noxious analytical manner of the Jews had been replaced with a good, clean, German synthetic approach. He was quite taken aback when the old man replied, with a mixture of deep sadness and anger: "Suffered? It hasn't suffered. It simply doesn't exist any more!"

Hilbert's life was dedicated to the search for knowledge. He abhorred the dispirited pessimism of the then-popular philosopher Emil DuBois-Reymond, who claimed that some problems were unsolvable, even in principle. The Frenchman's discouraging outlook was epitomized in the motto *"Ignoramus et ignorabimus"* ("We don't know and we shall never know"). Hilbert did not allow such pessimism to take over. When he died in 1943, his gravestone was inscribed with the words he once used in a radio lecture. They expressed his everlasting optimism: *"Wir müssen wissen, wir werden wissen"* ("We must know, we shall know").

At the turn of the twentieth century, the Frenchman Henri Poincaré could also lay down a legitimate claim to the title of most important mathematician of his time. But the two men were not rivals. In fact, they held

David Hilbert

each other in high esteem, and when Poincaré was chosen to organize the Second International Congress of Mathematicians, he asked Hilbert to prepare an address that would provide a memorable experience for all attendees. The congresses are held every four years, the first one having taken place in Zürich, in 1897. The second one was going to be held in the French capital, in 1900. Adding four to 1897 doesn't exactly come out to 1900, as any mathematician of Poincaré's stature could have told you, but the World Exhibition was simultaneously taking place in Paris, and that was too good an opportunity to pass up. In fact, it was the reason why about two hundred other scientific conferences were also held in Paris during that year.

When Hilbert received Poincaré's invitation, he hesitated at first. It would be very tempting to end the old century by giving a speech that would influence his colleagues and successors throughout the new one, but how could he capture the attention of the international audience that was expected at the congress? Then he thought of an interesting angle: maybe he could focus the development of mathematical research in the twentieth century by directing his listeners toward some important, unsolved problems. Before he made a decision, however, he took advice from his friends Minkowski and Hurwicz in Zürich. They endorsed the idea enthusiastically, but Hilbert continued to vacillate until it was nearly too late. The invitations for the congress had been sent out in June and did not list any speech by the German professor. But Hilbert had been writing all the time and in mid-July his speech was ready. Unfortunately, it was no longer possible to schedule it as the keynote speech, as Poincaré had planned, and his talk had to be transferred to the joint session for Bibliography and History, and Teaching and Methods. These sessions were regarded as quite inferior to the sessions on theoretical subjects, but Hilbert's address gave luster to these somewhat boring seminars.

Before Hilbert's lecture, some of the 253 attendees had expressed dissatisfaction with the limited social program that was prepared for them by the organizing committee. Some of the honorable mathematicians must have heard rumors about debauchery in chic Paris, and traveled from far and wide expecting some action. After all, the Folies Bergères had a world-famous show that was already in its thirty-first season. And the Moulin Rouge, now in its eleventh season, was famed for its cancan dancers, *"qui lancent leur jambe en l'air avec une élasticité qui nous laisse présager d'une souplesse morale au moins égale"* ("who throw a leg in the air with an agility that lets us suspect an equal moral suppleness"). Tickets to one of these spectacles—all in the interest of science, of course—would have been greatly appreciated. Other congress participants, more interested in sport than "folk dancing," would have been overjoyed to attend at least some of the events of the second Olympic games that were being held in the French capital at precisely the same period.[3] Even a visit to the World Fair or a climb up the Eiffel Tower would have been a welcome diversion.

So much to do, but nothing doing. It was mathematics from morning to evening and discontent started to rise among the participants. But all the mutterings were put to an end with Hilbert's lecture. "The Paris Congress will forever bask in a special glory," reads the official history of the International Mathematical Union. Forget the Moulin Rouge, the Eiffel Tower, and the Olympics. The mathematical congress "will be remembered in the history of mathematics for David Hilbert's address."

At the lecture, Hilbert immediately caught the attention of his auditors with his opening sentence: "Who among us would not like to lift the veil behind which the future lies hidden, in order to cast a glance at the

---

[3] They wouldn't have liked it; it was all chaos and confusion. Baron De Coubertin, the founder of the modern Olympic games, would say later: "It's a miracle the Olympic movement survived these Games."

advances that await our science, and at the secrets of its development during the coming centuries?" The professor expounded his vision for nearly an hour, and then turned to the unsolved problems. But he had already taken up most of the time allotted to his address and could not present all twenty-three problems. He made do with ten and had the full list distributed to the participants later on. His address ended with the conviction that "mathematics is the foundation of all exact knowledge of natural phenomena," and expressed the hope that "the new century bring it gifted masters and many zealous and enthusiastic disciples, so that it may completely fulfill this high mission."

Hilbert's speech was highly influential. It provided new impetus for research and spawned wholly new disciplines. One could say that Hilbert's address turned out to be truly prophetic, since it determined the direction of much of mathematical research for the next century. On the other hand, it was a rather self-fulfilling prophecy, since the address was meant to influence future mathematical research.

The list of problems encompassed all areas of mathematics. Hilbert, and with him Poincaré, were probably the last people who were able to grasp the entirety of mathematics. With the new century, research split into various disciplines and subdisciplines, whose exponents nowadays, more often than not, don't understand each other.

Not all problems on Hilbert's list have been solved yet. Problem 8, for example, deals with the Riemann hypothesis, which has been called the most famous, and is certainly the most important, unsolved problem in mathematics. It concerns the so-called Zeta function. Much depends on Riemann's hypothesis, because hundreds of principles, postulates, and theorems assume its validity—pending final settlement of the question—and base their own conclusions on this assumption. It may be a good thing it has not yet been solved. While a solution for any of Hilbert's unsolved problems promised no more than eternal glory, recently the ante has been upped. The Clay Mathematics Institute, a nonprofit institute dedicated to increasing and disseminating mathematical knowledge and set up by the Boston businessman Landon T. Clay, announced a $1 million prize for anyone who proves or disproves the Riemann hypothesis.[4]

Some of Hilbert's problems were solved, but differently than the professor expected. In Problem 1, Hilbert asked the community to prove that the objects of a set are either countable, by integer numbers, or are as dense as the continuum of real numbers. The situation can be visualized in the following

---

[4] Six other "millennium" problems were announced at the same time, each of which carried a prize of $1 million.

way: draw a line and count all the points on it. Of course, there are infinitely many points. Then cut out a section from this line and count the points on this section. Again, the number is infinite. The question is now: Is the infinite number of points on the section equal to the infinite number of points on the whole line? Or is the number of points in the smaller section "countable," that is, somewhat less than a continuum?

Hilbert illustrated the notion of infinity with an ingenious example. He told his listeners to imagine a hotel with an infinite number of rooms. All rooms are taken, but the manager neglected to hang the "no vacancy" sign. In the middle of the night a traveler arrives and asks for accommodation. "No problem," says the manager, and has all guests move from room number N to room number $N + 1$. The traveler can have room number 1. This shows that infinity plus one equals infinity. But wait, there's more. The new guest has barely settled into his room, when a bus arrives with an infinite number of passengers. Of course, the "no vacancy" sign still hasn't been put up, and with good reason. Can all the newcomers be accommodated? "No problem," says the manager, and has all the guests move from room number N to room number $2N$. With this move, an infinite number of odd-numbered rooms become available for the infinite number of new guests. So two times infinity is also equal to infinity. (By the way, each one of the newcomers gave the manager an infinitely small tip. How much did he receive in total compensation?)

In 1963, it turned out that Hilbert had posed the question for Problem 1 incorrectly. It is not *either/or*. One answer and its opposite are true. The mathematician Paul Cohen from Stanford University showed that some other notion of cardinality must exist between countability and continuum.

And in Problem 10 Hilbert asked his colleagues and the coming generations of mathematicians to devise an algorithm that would be able to determine whether a Diophantine equation has a solution.[5] No can do, said Yuri Matiyasevich, from St. Petersburg (then Leningrad) in 1970. In his dissertation he proved that such an algorithm cannot exist.

At least one problem never will be solved, strange as that may sound. It is Problem 2 on Hilbert's list, which deals with the consistency of arithmetic, that is, with its freedom from contradictions. We would like to think that arithmetic, as we know it, needs no further proof. Two plus two equals four; there are no two ways about it. But not only was Hilbert right in asking the question, worse news was to come. Thirty years after the congress, the logician Kurt Gödel showed that arithmetic is *not* free from contradic-

---

[5] A Diophantine equation is an equation whose coefficients are integers, like the one in Fermat's famous theorem.

tions. Both a statement and its opposite may be true simultaneously. Propositions could be formulated that are undecidable.

The standard example of an undecidable question is the following: If the barber of Seville shaves all the men of Seville who do not shave themselves, does he shave himself? If he does, he doesn't, and if he doesn't, he does. The question simply cannot be answered. Or consider this more exciting example. Let's say you land on an island inhabited by cannibals. You are caught immediately. The cannibals' chef prepares the menu and, in a jocular mood, lets you decide what dish you want to be served as. To spice up the preparations, he devises the following game: you may make a final statement. If it is true, you will be sautéed in a pan. If it is false, you will be grilled and served as barbecue. But maybe there is a statement that will save your life. There is, and please commit it to memory immediately, in case you should ever find yourself in such an uncomfortable situation. The statement is, "I will be grilled!" Once you utter that sentence, the chef's menu plans go down the drain. Is the statement true or false? If he grills you, you have told the truth, and you should have been sautéed. But if he sautés you, you have lied, and consequently you should have been grilled. There is nothing he can do: his buddies will have to go without dinner.

That was quite a blow to Hilbert. It threw everything he believed in into disarray. The lack of contradictions is, after all, the basis of a rigorous foundation of all of mathematics. But for once, mathematicians decided to take a pragmatic stand. They decided that mathematics is *probably* free of contradictions, but that this can't be proved in a finite number of steps. That's a copout if ever there was one, but if it's good enough for Hilbert and his followers, it is good enough for us. And so the game may continue.

Then there is Problem 18 on Hilbert's list. It went under the title "The building up of space from congruent polyhedra," and belongs to the so-called *specific* problems on Hilbert's list, in contrast to the more basic *general* problems. Hilbert divided the problem into three parts, one of which will be recognized by the astute reader as Kepler's problem.

In the first part he asks for a proof that a finite number of objects exist that can build up space, that is, that can tile it without gaps or overlaps. For two- and three-dimensional space, the question had already been answered in the affirmative before Hilbert raised it. There are exactly 17 different plane symmetry groups, and exactly 219 three-dimensional symmetry groups. The result for the two-dimensional case is of paramount importance for printing shops that produce wallpaper. It states that there are exactly seventeen distinct ways to print wallpaper in a pattern that is periodic in two directions. So don't let the salesperson in the home improvement store confuse you. Even though wallpaper can be decorated with an infinite number of flowers or ornaments or anything else, there are not more than seventeen symmetries (translations, rotations, reflections, or any combinations thereof). The answer for the three-dimensional case has even more important implications. It means that atoms can configure themselves into exactly 219 different molecules to form crystals. It is nothing less than astounding that such down-to-earth results were arrived at purely through mathematical reasoning, without recourse to experiments or observations.

The answer to the general, *n*-dimensional case was found in 1910 by Ludwig Bieberbach. He proved that, not only in two and in three dimensions but in *all* dimensions, there are a finite number of objects that can build up space. The ambitious young man, an ardent German nationalist and later a prominent Nazi, was only too happy to solve one of Hilbert's celebrated problems—even if it was only a subproblem. It was to be an important first step on his career path. But his later association with the Göttingen professor became very strained. In 1928 Bieberbach insisted on boycotting a mathematicians' congress for nationalistic reasons, because it was to be held in Italy. Hilbert fumed. "We are convinced that Herr Bieberbach's way will bring misfortune to German science and will expose us to all justifiable criticism from well disposed sides," he wrote to colleagues around the country. Hilbert's view prevailed, and in Bologna, Italy, the Germans formed the second-largest national contingent, after the Italian hosts. With Hitler's rise to power, Bieberbach became a leading anti-Semite, wore a Nazi uniform when conducting examinations, and took a leading role in dismissing Jewish colleagues from their positions. After the end of World War II, he was stripped of all his posts.

Hilbert gave the second part of the problem a slightly different tilt. He asked whether objects exist that (1) fill *n*-dimensional space without gaps and without overlaps, but that (2) cannot be transformed into each other by simple movements. The first requirement again means that the objects must tile the space. The second requirement means that all tiles have to be exact copies of each other, but cannot be brought to cover each other just by sliding them around in different directions. In other words, they must be somehow similar but not identical.

It took twenty-eight years to provide a first answer. In a 1928 paper presented to the *Akademie der Wissenschaften* (Academy of Sciences) in Berlin, Karl Reinhardt described a complicated three-dimensional object, copies of which completely fill three-dimensional space—without gaps and without overlaps—but which cannot be brought to cover themselves by simply moving them around. He also formulated the conjecture that no such tiles exist in two dimensions, using the same method that Ludwig

August Seeber employed for his conjecture on ternary quadratic forms. He spent lots of time trying to find an appropriate tile. After he had tried and tried without success, the exasperated mathematician convinced himself that no such shape could exist. Someone else would certainly provide the proof and his name would forever be connected to the conjecture-turned-theorem.

That's a really bad way of doing mathematics, and it came as it had to. Barely four years later, Reinhardt was proved wrong. The German violinist and mathematician Heinrich Heesch found a shape that can tile the floor, but whose representatives cannot be brought to exactly cover each other by sliding them around. Some of the tiles could only be covered if another tile is lifted off the floor and flipped around. When the venerable German ceramics company Villeroy & Boch got wind of Heesch's discovery, they saw an opportunity to branch out into nonconventional floor coverings and immediately offered to produce such tiles. Heesch designed a suitable shape, which was then used for the ceiling of the Göttingen town library, where they can still be seen today.

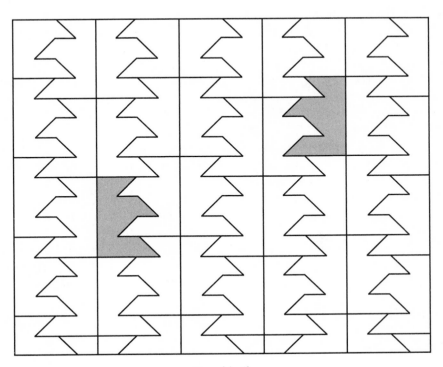

Heesch's tile

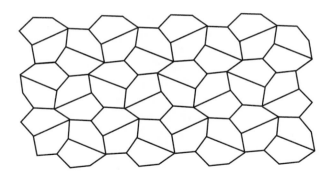

Kershner's pentagonal tile

Some readers may have expected a convex tile, while Heesch's shape is concave.[6] But even if we were to rule out concave shapes, something Hilbert did not require, Reinhardt's conjecture is false. Richard Kershner (whom we met in chapter 4) found a pentagonal shape in 1968 that (1) can fill a kitchen floor without gaps and overlaps, (2) has tiles that cannot be transformed into each other by movement alone, and (3) is convex. From all this we learn an extremely important lesson. If you absolutely must propose a conjecture, at least make sure that you won't be proven wrong for a long, long time—if possible, not during your lifetime.

But it is the third part of Problem 18 that we are interested in. Specifically, Hilbert's question was: "How can one arrange an infinite number of equal solids, of given form, most densely in space, e.g., spheres with given radii. . . . [H]ow can one fit them together in a manner such that the ratio of the filled space to the unfilled space be as great as possible?" That is, Kepler's conjecture as tossed into the twentieth century. Hilbert mentions that an answer to this question is "important to number theory and [may be] sometimes useful to physics and chemistry."

The race was on again and mathematicians went back to work. It soon became apparent that a proof was still way beyond reach. The available mathematical tools just did not suffice to crack the problem. Try as they might, nobody even got close to solving Kepler's conjecture. So the mathematics community set itself a more moderate, intermediate, goal: find upper and lower bounds for the best density. This is a time-tested method in mathematics whenever the exact answer to a problem is not known. Establish upper and lower bounds and show that the true result must lie

---

[6]Concave and convex are the mathematical terms for bulging in and out.

somewhere between these two limits; then, try to narrow the gap by raising the lower bound and lowering the upper bound. As the gap gets compressed, claustrophobia sets in as the result gets caught between ever more cramped walls. What happens when the upper bound has been lowered all the way down to the lower bound, or vice versa? The result gets locked in, and the exact result has been found.[7]

Beware, however: if the lower bound turns out to be higher than the upper bound, there's trouble. No result can be smaller than the lower number and, at the same time, greater than the higher number. (You may take some solace, however, if this happens: getting the bounds to cross is a convenient way to show that no result exists.)

As we saw in the appendix to chapter 1 and at the end of chapter 7, the arrangement suggested by Kepler has a density of 74.05. Does an arrangement exist with a density higher than that? If it can be shown that no packing can have a density greater than 74.05 percent, then Kepler's arrangement of spheres is the densest packing possible. It may not be the *only* such packing, because there may be others with the same density. In fact, as William Barlow discovered, there is an infinite number of such packings (see chapter 1). So the accepted strategy at that time was to narrow the gap between the upper and the lower bounds.

The density of the best packing must be at least as high as the density of Kepler's arrangement. So 74.05 percent constitutes a *lower* bound. On the other hand, nothing can be stuffed into a space once it is full, especially not spheres. So 100 percent is the highest conceivable density in any space and this number constitutes an *upper* bound. The density of the best sphere packing must, therefore, lie somewhere in between these two bounds. If a lower bound could be established that is greater than 74.05 percent, Kepler would be proven wrong, and his would not be the densest arrangement. In such a case, Kepler's sphere arrangement would still represent the best *lattice* packing, since Gauss had conclusively proved that, but some jumbled up arrangement would be denser still. But nobody took this possibility seriously and no mathematician wasted his time trying to raise the lower bound. Instead, all efforts were directed at lowering the upper bound little by little, in the hope that eventually an upper bound of 74.05 percent would be reached. This task was to keep mathematicians busy for most of the twentieth century.

---

[7] This is the method Odlyzko and Sloane employed when they derived the highest kissing number in eight and twenty-four dimensions (see chapter 6).

CHAPTER 9

The Race for the Upper Bound

The first to try his hand at establishing an upper bound well below 100 percent was Hans Frederik Blichfeldt. This man's life story is one of those fairy tales that relate an ascent from ashes to glory. Blichfeldt managed to rise from the ranks of a simple laborer to become head of the mathematics department of one of the world's most prestigious universities. He was born in 1873 in Denmark, the son of a farmer. After the family's emigration to the United States, the fifteen-year-old boy worked as a farmhand and laborer in sawmills and lumber companies. But he did not intend to stay a workman all his life. After a few years, he became a draftsman, then a surveyor, and finally entered Stanford University. He spent his academic career at Stanford, starting as an instructor and working his way up through the ranks—lecturer, assistant professor, associate professor, and finally full professor. In 1927 he was appointed head of the mathematics department. In this post he managed to make Stanford one of the foremost centers for mathematics of the world.

In 1919 and 1929 Blichfeldt published two papers concerning sphere packings. An ingenious idea allowed him to establish the first upper bounds for the packing density of three-dimensional spheres below 100 percent.

Imagine a large box and, within this box, a jumbled-up mess of spheres. Now increase the radius of the spheres and, at the same time, enlarge the box so that there is sufficient room in it for the larger spheres. Parts of the enlarged spheres may overlap. Blichfeldt's idea was to fill the spheres with layers of sand, using a sophisticated method. The layers become thinner the farther away one gets from the center. So there would be lots of sand in the middle of a given sphere, but as one approached the outside surface there would be less and less.

You could imagine the spheres as onion-like objects, whose layers in the center are dense and heavy and then become increasingly thin and light the farther one moves from the center. Since some spheres overlap, the sand of different spheres combine in the regions of overlap. Here comes Blichfeldt's clincher: by judicious distribution of the sand—and Blichfeldt developed formulas for this— it can be guaranteed that nowhere will there be more sand than in the dense center of the spheres. This will be true even if several spheres overlap, due to the sand's sparseness toward the outer surfaces.

The rest is easy. Weigh the box that contains the sand-filled spheres and compare the result with the weight of an identical box that is completely filled with sand. Obviously, the latter contains more sand and is, therefore, heavier than the former, no matter how many spheres it contains. Hence, when you divide the weight of the box that is completely filled with sand by the weight of the box that contains the spheres, you get a number that is less than 1.0. Moreover, this ratio represents an upper bound on the density of spheres in a box. Now let the boxes' sizes grow very, very large and compute the ratio again. The numerical value turns out to be 0.883. Whatever the arrangement of spheres within an infinitely large box, their density can never exceed 88.3 percent.

Compared to 74.05 percent, the density of Kepler's arrangement, an upper bound of 88.3 percent still leaves a lot of leeway. A denser arrangement of spheres is, in principle, not ruled out, but Blichfeldt did narrow down the possibilities.

In 1929, he followed up with an improvement. The crucial point in his derivation of the upper bound was the way in which the sand was distributed within the spheres. In his second paper, he devised a different way to fill the spheres with sand and managed to lower the upper bound from 88.3 percent to 84.3 percent. Moreover, at the end of the paper he declared that "a more elaborate use of the relations" would lower the upper bound even further. Without further ado, he claimed that "in no case can equal spheres be packed . . . such that the space occupied by the spheres is as much as $^{835}/_{1000}$ of the volume of the cube."

We now have an upper bound of 83.5 percent. By filling spheres with sand, figuratively speaking, Blichfeldt had made the first steps towards solving Kepler's conjecture since Gauss's book review a century earlier. With clever reasoning, he established that no packing could exist with a density higher than 83.5 percent. But an arrangement with a density up to that number could, in principle, exist, and if one could be found it would spell the death of Kepler's conjecture. But Blichfeldt's proof gave no indication what such a packing would look like and, pending further evidence, Kepler's conjecture remained undecided.

From here on the going became tough. For eighteen years Blichfeldt was the holder of the world record: his 83.5 percent remained the best available upper bound until 1947. Then a Scottish mathematician entered the picture, Robert A. Rankin. Rankin was born in 1915 in Garlieston, Wigtonshire. During the war he worked on rocket flight. At the same time, across the Atlantic, his colleague Richard Kershner was working on similar problems (see chapter 4). Stints at Cambridge and Birmingham followed. In 1954 he joined the University of Glasgow as a professor of mathematics, and stayed there for twenty-eight years. From 1971 to 1978 he served as clerk of the university's senate, and for many years was head of its mathematics department. Even after his retirement, in 1982, he stayed active in research. In 1987 he received the Senior Whitehead Prize. The prize is bestowed by the London Mathematical Society on worthy mathematicians in every odd-numbered year. (The Junior Whitehead Prize is awarded to mathematicians under the age of forty.) In 1995 Rankin received the De Morgan medal, the London Mathematical Society's most prestigious prize. It is awarded to worthy mathematicians in years that are divisible by 3. Fortunately, Rankin had already received the Senior Whitehead Prize, because the rules for the latter state that it may not be awarded to any person who has previously received the De Morgan medal. (Very puzzling indeed, these British rules.)

Toward the end of 1946, Rankin, just released from war duty, found himself at Clare College in Cambridge. That's where he wrote a twenty-page paper on an upper bound for the sphere packing. After a rather dull start, in which Blichfeldt's derivation is rehashed, complete with proofs, Rankin took off. He did not just seek a better method for the distribution of sand, he sought the best. And he found it, by trial and error. One of the difficult parts of the proof was the computation of the numerical values of a certain formula. In 1946 such a calculation was still a major undertaking, although it would take only a few seconds today using a hand calculator. Fortunately, the Cambridge Mathematical Laboratory came to the rescue and "provided a calculating machine by means of which the computations of the numbers were carried out." Once that hurdle was overcome, Rankin could, again, weigh the sand-filled spheres, and establish a new upper limit: 82.7 percent.

This is as far as Blichfeldt's sand-filling method could advance. Rankin had exploited it to its maximum; no further improvement was possible. To move matters ahead, one would have to travel down a totally different avenue.

This is what the Hungarian mathematician Laszlo Fejes-Tóth tried to do in 1943. Fejes-Tóth had already shown his mettle three years earlier, with

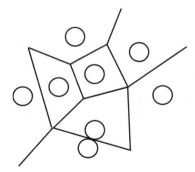

V-cells in two dimensions

the proof of the sphere packing problem in two dimensions (see chapter 4). The invaluable experience that he gained there made him uniquely qualified to push the upper limit down. Leaving the physical aspects of sandfilled spheres behind, Fejes-Tóth returned to the fundamental geometrical properties of the sphere packing problem. In particular, he studied the empty area in the vicinity of each sphere. His idea was to divide space into cells, such that each sphere would sit in its own cell. Like good neighbors, the spheres divide up the area between them equally. Hence, the walls run exactly halfway between the spheres. If two spheres touch, the cell's wall goes through the point of contact.[1]

These cells, which will play a fundamental role in the further history of Kepler's conjecture, were initially proposed by the mathematician Georgii Feodosevich Voronoi. The son of a professor of Russian literature, Voronoi was born in 1868 in Zhuravka, a town that was then in Russia and now belongs to the Ukraine. He studied physics and mathematics at the University of St. Petersburg, where both his master's thesis and his doctoral dissertation were awarded prizes. Then he joined the faculty of the University of Warsaw. In 1904 he attended the Third International Congress of Mathematicians in Heidelberg and met Hermann Minkowski. To their great surprise, the two men discovered that, unbeknownst to each other, they had both been working on the same subject: the geometry of numbers. Unfortunately, there was no occasion for closer collaboration, because Voronoi died four years later, at the age of forty. He left an important legacy, however: three long papers on quadratic forms. Another legacy is represented by the Voronoi cells, or V-cells for short, which will help us find improved upper bounds for the packing density. If all spheres sit inside identical V-cells, then the smaller the V-cell that encloses a sphere, the higher the packing density.

---

[1] We met Voronoi cells in chapter 4. They were the gardens on the oceanfront property.

When Fejes-Tóth decided to inspect V-cells, he thought a good candidate would be the regular dodecahedron. The smallest dodecahedron that encloses a sphere of radius 1 has a volume of 5.5503 (see the appendix for some details of how to compute this volume). Since the sphere has a volume of 4.1888, this would establish an upper bound for the density of $^{4.1888}\!/_{5.5503} =$ 75.46 percent. Wait a minute—this is greater than the density of Kepler's arrangement (74.05 percent), which we believed was the best one possible. Is it possible that Fejes-Tóth had found a better packing arrangement?

The answer is an emphatic *no,* and the reason is simple. Even though the dodecahedron does give a tight fit locally, the arrangement cannot be extended to several spheres. When you try to fit dodecahedra together, gaps appear. Three-dimensional space cannot be completely filled with dodecahedra.

It is ever so unfortunate that not every object that is regular locally stays regular globally. Take Platonic solids, the regular polyhedra in three-dimensional space. Platonic solids are structures that fulfill two conditions: (1) all faces of the solid are identical, regular polygons, and (2) the same number of faces meet at each vertex. The cube, for example, is composed of six squares, and three of them meet at every corner. Only five Platonic solids exist: the tetrahedron (four faces), the hexahedron or cube (six faces), the octahedron (eight faces), the dodecahedron (twelve faces), and the icosahedron (twenty faces).[2] Let's inspect the dodecahedron. Taken on its own it is quite regular; all its faces are equal-sided pentagons, and three pentagons meet at every vertex. But it all depends on what your definition of *regular* is. Dodecahedra do not tile space!

This unfortunate fact of dodecahedral life was discovered in the 1980s, to the chagrin of the computer scientist-turned-sculptor Robert Dewar. Dewar, who works in Tehachapi, north of Los Angeles, is inspired in his art by the subtle secrets of nature, for example, by the strange symmetries of polyhedra or by the structure of molecules. An artist and not a mathematician, Dewar once tried to pack twelve dodecahedra around a central dodecahedron. Dodecahedra, after all, are regular polyhedra in the same way as hexagons are regular polygons. Hence what works in the plane should work in space, he thought. But the dodecahedra would not fit. They could be almost joined together, but never exactly. A small, wedge-shaped space was always left open between the faces of adjacent dodecahedra. Exasperated, he

---

[2] Note that the tetrahedron is not a pyramid, even though it is similar. Pyramids, as found in Egypt, have a square base, while the remaining faces are triangles. The pyramid therefore belongs to the group of semiregular polyhedra, or Archimedean solids. Two pyramids, placed base to base, form an octahedron.

turned to a simpler task and attempted to sculpt a ring of dodecahedra by stringing six of them together end to end, leaving a hexagon open in the middle. Again, he didn't succeed: after placing the first five dodecahedra, there was not enough room left for the sixth. Soon the reason why nothing seemed to fit became obvious to Dewar. The dihedral angle (the angle of intersection of any two adjacent faces) of a regular dodecahedron is 116° 34'. Three of these angles add up to 349° 42', which is just 10° 18' short of the 360° required to close the circle,[3] hence the gaps. For comparison, the angles of the hexagon are each 120° wide, so three of them combine to a full circle. The same holds for four squares (with 90° corners), and six triangles (with 60° corners).[4]

If dodecahedra can't be joined together, maybe they could at least provide a new upper bound for the packing density. Fejes-Tóth thought they could, and set out to prove that the V-cells of any sphere packing have to be at least as large as dodecahedra. In a 1943 paper in the *Mathematische Zeitschrift,* he proved that honest-to-goodness V-cells must have volumes of *at least* 5.5503. Therefore, the packing's density would have to be smaller than 75.46 percent. His approach provided an upper bound that was significantly lower than both Blichfeldt's and Rankin's. There was only one problem: the proof contained mistakes. They weren't spotted immediately, and Fejes-Tóth bathed in glory for a while. But the errors were detected a few years later.

Where did Fejes-Tóth go wrong? Most of the proof had been worked out very neatly. But on the next-to-last page, just before the end of the proof, he made two seemingly harmless assumptions, one of which turned out to be fatal. The first assumption was that no more than twelve spheres can simultaneously touch a central sphere. Fejes-Tóth did remark, in footnote number 6, that proving this assertion is no easy feat, but then left matters at that. We'll let that one pass since we know—with hindsight—that Schütte and van der Waerden did provide a correct proof for the kissing problem ten years later, in 1953. It was a bad start, however.

Fejes-Tóth next investigated how close a thirteenth sphere could get to the central sphere (all of radius 1). Without much ado, he stated that it can't

---

[3] To visualize, the large hand on your watch requires one minute and forty-three seconds to cover an angle of 10° 18'.

[4] Remembering his original calling as a computer scientist, and inspired by a book on crystallography, Dewar developed a computer program that slightly deforms the faces of the dodecahedra, until they—now no longer regular objects—fit together. The deformations can hardly be seen by the naked eye. Incidentally, the only Platonic solid that completely fills three-dimensional space is the cube.

Dodecahedron as a V-cell

get closer than 1.26. He justified this number in footnote number 7 with the explanation that a rough experiment had convinced him that the distance between the central sphere and the thirteenth sphere must actually be larger than about 1.38. So 1.26 certainly seemed an okay assumption. *A rough experiment?* Had he not learned anything from David Hilbert's uncompromising, rigorous approach to mathematics? No mathematician is ever allowed to formulate a proof that is built on the basis of a rough experiment. Did he really hold a central sphere in his fingers, juggle a dozen spheres around, and then put a ruler to the thirteenth sphere, to measure its distance from the central sphere?

When he tried to mend the holes, the proof of this seemingly innocent assumption resisted all efforts. After years of trying to fill the gap, Fejes-Tóth finally gave up. To his credit, it must be said that he himself admitted defeat and classified the dodecahedral conjecture an open problem in 1964. The wording of the conjecture was: "In any unit sphere packing, the volume of any Voronoi cell around any sphere is at least as large as a regular dodecahedron of inradius 1." The dodecahedral conjecture, which is obviously related to, but doesn't solve, Kepler's conjecture, has in fact been solved, but we'll talk about that in chapter 14.

So it was back to square one. Fifteen years after Fejes-Tóth's attempt, in 1958, a new approach was provided by the British mathematician Claude Ambrose Rogers. Rogers, who throughout his career made contributions of fundamental importance to the problems of packing and covering, was born in Cambridge, England, in 1920. He attended a boarding school and, upon turning eighteen, went to study mathematics at University College in London. He received his B.A. degree in 1941, but the war was still on, and he was called to serve his country in the Ministry of Supply. At the end of the war, he rejoined University College and, while serving as a lecturer, worked his way through a doctorate. A four-year stint at the University of Birmingham followed. Finally, he was called back to his alma mater, University College, as Aston Professor of Pure Mathematics. There he stayed until his retirement in 1986. He wrote over 170 research papers and in 1977 the London Mathematical Society awarded him the De Morgan medal. That was eighteen years before Rankin won his medal.

But Rogers gained an even higher distinction: he carries the coveted Erdös number ½. This number indicates a relationship (in terms of coauthorship) to the most prolific mathematician of all time, the Hungarian mathematician Paul Erdös. With about fourteen hundred published papers, Erdös, who died in 1996, is the world record holder in mathematical output. (The second-ranked author has about half as many papers to his credit.) But Erdös was no solitary scientist, jealously keeping his ideas to himself. Like no other mathematician, he promoted cooperation among his colleagues. He traveled the world, staying with anybody who was willing to have him—it is reported that he had no home to call his own—and doing research with his eager hosts. He coauthored papers with an unprecedented number of mathematicians.[5] And this is how Erdös numbers came into being.

An Erdös number of 1 is awarded to mathematicians who wrote at least one paper with Erdös. At the end of the year 2000, there were no less than 507 mathematicians who proudly carried an Erdös number 1. And as of 2002 the list had not been finalized, since new papers were still being added to his list of publications. (Even a few years after his death, his coauthors were still revising and correcting work that was done in collaboration with him.) By the way, there is also a famous nonmathematician with an Erdös number of 1: the baseball player Henry L. "Hank" Aaron autographed the same baseball as Erdös when they were both getting honorary degrees from Emory University in 1995.

Erdös's 507 collaborators have, themselves, 5,897 coauthors. These people carry an Erdös number of 2. Then there are 26,422 people who collaborated with mathematicians with an Erdös number of 2 and who, therefore, have an Erdös number of 3. And so on. Nearly every research mathematician today has an Erdös number of 6 or lower, and only 2 percent have Erdös numbers higher than 8.

As the number of mathematicians with Erdös number 1 increased, the need arose to distinguish among those relatively few colleagues who wrote more than one paper with the master—there are two hundred of them. It was decided that they would be awarded a fractional Erdös number below 1,

---

[5] Contrast Erdös to Andrew Wiles, who worked in solitary for many years on Fermat's last theorem. Erdös very much disapproved of Wiles's reticence.

depending on the amount of work they had done with him. Rogers wrote seven papers with Erdös, whence derives his Erdös number of $\frac{1}{2}$. (There are fifty-one mathematicians with Erdös numbers smaller than, or equal to, $\frac{1}{2}$. The mathematician closest to zero is Andras Sarkozy, with an Erdös number of $\frac{1}{62}$.)

Back to Rogers. A large part of his career was devoted to Kepler's conjecture and related problems. His book *Packing and Covering,* published in 1964, summed up the current state of the art. In his 1958 paper we find a quote that has become famous among packing experts: "Many mathematicians believe, and all physicists know" that Kepler's conjecture is true. His quote expresses in a nutshell the frustration the mathematical community felt at the state of affairs. It was obvious to all, be they grocers, cannonball stackers, or physicists, that Kepler's arrangement of spheres was the densest one. But for three and a half centuries mathematicians have not been able to come up with a proof. Rogers's words also summarize the different outlook between the professions. Mathematicians require proof for every single, ever-so-obvious detail, while to physicists, something as well established as Kepler's conjecture requires no further proof. In 1976, John Milnor, then at the Institute of Advanced Study at Princeton, commented on the state of affairs by exclaiming, "the problem in three dimensions remains unsolved. This is a scandalous situation since the presumably correct answer has been known since the times of Gauss." And then he added, as an afterthought, a small detail: "All that is missing is a proof."

In 1958, while at the University of Birmingham, Rogers published a paper in which he derived an upper bound for the density of sphere packings that was just a tad under 78 percent. To be exact, he managed to lower the upper bound to 77.97 percent. Rogers inspected a tetrahedron and placed a sphere at each of its four corners. He then showed that even the best packing, whatever it would turn out to be, could not be denser than this configuration.

Some people claim that by reverting back to geometrical tools, Rogers's derivation made the intricacies of sphere packings more transparent. After all, Blichfeldt's method had more to do with the shovelling of sand into balls than with geometry. Rogers brought the subject down to the realm of geometry. But transparency, like beauty, is solely in the eyes of the beholder, and the average beholder will find little that is transparent in Rogers's paper—pages and pages, twelve in all, of equations, formulas, and integrals, without even a tiny illustration to lighten the fare. It would have been easy to clarify what Rogers meant by the following sentence, for example, if he had only resorted to a drawing: "To each center c of a sphere of the system, we assign the set Π (c) of all points of space, whose distance from c is equal to their minimum distance from the centres of the spheres of the system." As the picture shows, Rogers was describing a V-cell.

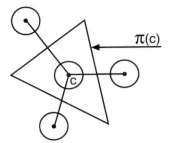

$\pi(c)$

Rogers's sentence

Following Fejes-Tóth's first, unsuccessful attempt, Rogers also decided to partition space into V-cells, each of which contained one ball. No gaps were left over in between. But he went a step further. He took a big kitchen knife and cut each and every V-cell into little pieces. In mathematically more appropriate language, he dissected the V-cells into simplices. A simplex is a little pyramid whose top is at the center of the V-cell, and whose base is one of the outside walls of the V-cell.

Each simplex contains part of a sphere; Rogers sought an upper bound for the density of a typical simplex. He argued that since the simplices make up a V-cell, and the V-cells completely fill space, it would suffice to average the densities. The resulting number would be an upper bound for the density of the whole space. Using this argument, Rogers went on to show that the density of the best sphere packing could not be greater than the density of four balls placed at the corners of a tetrahedron.

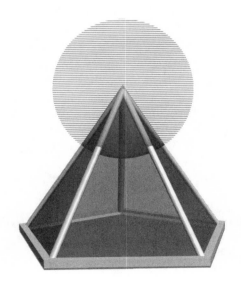

Simplex, with partial sphere

All that remained was to calculate the density of the balls in the tetrahedron. Such a computation is no picnic for regular mortals, but for an able mathematician it is a piece of cake. (The appendix shows how it is done.) The result: 77.97 percent of the tetrahedron is filled up by the four balls at its corners. This density is better than halfway between Blichfeldt's early attempts at an upper bound and the density of Kepler's sphere arrangement. The paper was a real advance and, henceforth, Rogers's suggestion to decompose V-cells into simplices became an important element in the study of sphere packings.

Let us compare the tetrahedra, with spheres at its corners, with Kepler's arrangement. How is it that the tetrahedra produce a density of 77.97 percent while Kepler's arrangement has a density of only 74.05 percent? Isn't Kepler's arrangement the densest possible? And doesn't it also place the balls at the corners of tetrahedra? The answer is that in Kepler's arrangement some of the balls are placed at the corners of tetrahedra and others are placed at the corners of pyramids with a square base. So spheres sometimes come to lie on top of three, sometimes on top of four, other spheres. A different way to recognize this is to realize that tetrahedra cannot fill space completely. Other objects, in this case square pyramids, are required to fill the gaps: each cluster of "12 around 1" consists of six pyramids and eight tetrahedra. (In the appendix I show that four balls in a tetrahedron fill 77.97 percent of the tetrahedron's volume and five balls in a square pyramid fill 72 percent of the pyramid's volume.) The density of Kepler's arrangement is a weighted average of the above densities.[6] By the way, this example shows again that locally a packing can be achieved that is denser than Kepler's arrangement. But in the global case, clusters of less density offset the dense clusters.

For many years Rogers's upper bound stood like a solitary beacon in the emptiness of the geometrical desert, guiding all those who came after. It was the benchmark against which any further progress had to be measured. For nearly three decades nobody succeeded in bettering the record. Then, in 1987, J. H. Lindsey II from Northern Illinois University managed to lower the upper bound even further. Lindsey was an outstanding student at Caltech; in 1963 he received the E. T. Bell Undergraduate Mathematics Research Prize—an annual cash award of $500 that is presented for the best original mathematics paper written by a Caltech junior or senior. Five

---

[6] The weighted average, with weights 6 and 8, gives the density of the dodecahedron: $^{(6 \times 72\% + 8 \times 77.97\%)}/_{14} = 75.46$ percent. In order to compute the density of Kepler's FCC arrangement, the weights are $6(\sqrt{3/2})$, and $8(\sqrt{1/8})$. The irrational numbers arise from the fact that the dodecahedron does not tile space.

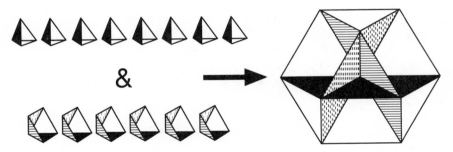

Kepler's arrangement consists of eight tetrahedra and six pyramids

hundred dollars may not seem like a lot of money today, but it meant a lot to an undergraduate forty years ago. He did his Ph.D. in group theory at Harvard University.

Lindsey's paper—the one on upper bounds, not the one for the prize—was one page shorter than Rogers's article, and had no pictures. I cannot resist but quote the following lines from his article: "Let A be the part of F bounded by a perpendicular from Q to the midpoint M on edge E, the line segment MV where V is one of the two vertices of E, and the line segment VQ. Let q, m, and v be the vectors from P to Q, M, and V, respectively." All clear? Not quite. Compare this mathspeak to the picture.

Clearer? I would hope so.

While Fejes-Tóth and Rogers had set the stage with the dissection of V-cells, Lindsey developed the method into a fine art. Fejes-Tóth used a chainsaw to separate space into V-cells, and Rogers used a kitchen knife to cut V-cells into pyramids. Lindsey applied a scalpel to the pyramids and carved them up even further. He divided each face of the V-cell into triangles,

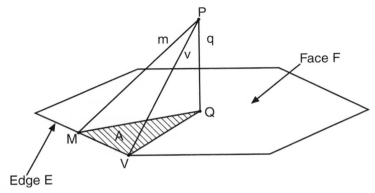

Lindsey's sentence

and constructed pyramids with these triangles as bases. Lindsey's pyramids were just as high as Rogers's, but had smaller bases. Then Lindsey computed the density in these pyramids, meticulously documenting each step of his calculations. Toward the end of his paper he got into a bit of a hurry, however, and decided to include most of the remaining arguments into a single statement:

> The angle VQW is twice the angle MQV so the angle MQY is 5 times the angle MQV so Y is represented by a real multiple of $((\tfrac{1}{8}^{\frac{1}{2}}) + (\tfrac{1}{6}^{\frac{1}{2}}))^5$ and of $(2^{\frac{1}{2}} + i)^5 = -(11)(2)^{\frac{1}{2}} + i$ which has smaller angle with the negative real axis than $2^{\frac{1}{2}} + i$ with the positive real axis, so when we put the regions for $i = 5$, 6 clockwise from that for $i = 4$ we get final vertex a real multiple of $-(11)(2)^{\frac{1}{2}} - i$ and in the angle between $-(11)(2)^{\frac{1}{2}} + i$ and $-(11)(2)^{\frac{1}{2}} - i$ only room for the region for $i = 1$.

Even a picture gallery won't help here. The long and the short of it is that the density of spheres in three-dimensional space cannot be greater than 0.7784. The upper limit is lowered to 77.84 percent.

After that, the pace picked up a little. Following Lindsey's paper, it took only one year for the mathematician Doug Muder to refine the method. But progress became so microscopic we have to move to five digits after the decimal point to spot the improvements. Muder was a mathematician working at MITRE, a nonprofit corporation in Bedford, Massachusetts, whose clients include the Internal Revenue Service, the Federal Aviation Authority, and the Department of Defense. Apart from designing methods for more efficient tax collection, the people at MITRE did a lot of cool things. The paper Muder submitted to the *Proceedings of the London Mathematical Society* in August 1986, for example, was called "Putting the best face on a Voronoi polyhedron." Now that's a title that invites readers to delve into the twenty-page paper. (They had to wait two years, however, until 1988, for its publication.) Muder, a nonconformist if there ever was one, broke with tradition: he included three illustrations!

He broke with tradition in other ways too. A friend describes him as "a mathematician whose career involved explaining complicated subjects to people who could fire him if they didn't understand. Until recently, his main works were impenetrable papers about such important topics as the precise number of Ping-Pong balls that can fit into a very big cardboard box. He finally cracked under the strain of taking all this seriously." Muder took time out to write computer books for idiots. Actually he was a bit more polite, and the books he wrote or co-authored carried titles like *Internet for Dummies* and *E-Mail for Dummies*.

Seventy years earlier, Blichfeldt had suggested the sand-filling method. Then Rankin had honed it to perfection. A generation later, Fejes-Tóth had suggested the V-cells method. Then Rankin had improved on it. Now it was up to Muder to hone that method to perfection.

You'll recall that the smaller the V-cell, the higher the density of the packing. So how can the volume of the V-cell be reduced? Muder thought that this could be achieved by taking a close look at the shape of the V-cell's faces. As the title of his paper indicates, he set himself the task of "putting the best face on a Voronoi polyhedron." After careful deliberation, Muder concluded that the most efficient face would have to be circular, small, and as close to the center of the V-cell as possible. From all the shapes Muder could think of, regular pentagons come as close to fulfilling these requirements as possible. Hence, he concluded, pentagonal shapes are the V-cell's best faces. So he dissected space into V-cells with pentagonal bases and cut them into pyramids. To distinguish his paper from Rogers's paper, Muder did not call the pyramids *pyramids;* he called them *wedges.* He then assembled them into wedge-clusters, and calculated their volumes. Comparing the wedge-clusters' volumes with the volume of a sphere, he managed to lower the upper bound for the packing density to 0.77836.

So the upper bound now stood at 77.836 percent. This is an improvement on Lindsey's bound of less than one hundredth of one percent. Both Blichfeldt's sand-filling method and Rogers's V-cell method had been manipulated with great skill, and no further progress seemed possible unless, of course, a new idea arose that would open up a different avenue.

Muder remained fascinated with the problem and spent the next few years thinking about the matter. He was all the more intrigued, since—according to rumor—Lindsey had been busy in the meantime, and had found an upper bound that was better than his: 77.36 percent. Soon, the news was no longer just a rumor. A working paper appeared and was distributed by Lindsey to his colleagues. But this upper bound never saw the light of day in any learned journal. We may assume that Lindsey had followed a false lead, and that his colleagues made him aware of that.

But the rumor did have a positive effect: it kept Muder going. How could the upper bound be lowered? Suddenly it hit him: pentagons may not be the best faces for the V-cell after all. They were the best *straight-edged* bases. But a good mathematician thinks outside the box. Perhaps the volume of the V-cells could be minimized further, if other face shapes were considered.

Now we are entering the realms of microsurgery. Muder started shaving, chiseling, trimming, and carving the bits and pieces into which the V-cells had been cut up. Throughout, he took care to sculpt a shape that would fill

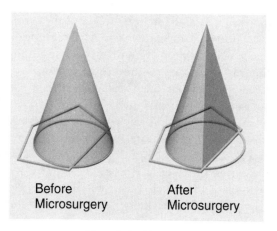

Before
Microsurgery

After
Microsurgery

Shaved circular cone

the space without gaps and, at the same time, leave sufficient room to contain the sphere. Finally the master craftsman was done. Volumes had been reduced to their skinniest minimum. What was left were thirteen *shaved circular cones,* that is, pyramids that initially had round bases, but from which some parts had been removed.

In order to make his point, Muder needed three propositions, which he proved with the help of no less than twenty subpropositions and *technical lemmas.* The V-cells had been cut into thirteen shaved circular cones whose solid angles were identical to the solid angles of the "best faces" from his previous paper. The crucial point was that the new and lean, if not to say emaciated, V-cell had a volume of only 5.41848. Since a sphere of radius 1 has a volume[7] of 4.18879, the density is $^{4.18879}/_{5.41848}$ or 77.306 percent. By skillful use of modeling tools, Muder had managed to lower the upper bound by a whopping one-half of one percent.

It had not been easy going, and the ordeal was far from over. Once he had everything neatly worked out, Muder wrote up an article and, on July 18, 1991, submitted it to the journal *Discrete and Computational Geometry.* It was not satisfactory and he was asked to revise his work. A year and a half later, in February 1993, Muder submitted a revision. His revised article was still not acceptable to the editors. Finally, in June of the same year, Muder's second revision was accepted for publication. The article appeared a few months later.

---

[7] Volume = $\frac{4}{3}\pi$.

Altogether, in the seventy-four years since Blichfeldt's first attempt, the upper bound was lowered to 0.77306. Progress had been agonizingly slow. With minute advances every dozen years or so, the upper bound inched downward like the world record in the 100-meter dash. On July 6, 1912, the American sprinter Donald Lippincott covered the distance in 10.6 seconds. On September 14, 2002, in Paris, American athlete Tim Montgomery lowered the world record to 9.78 seconds. The world record had improved by 7.8 percent over a period of ninety years. During about the same period, the upper bound for the packing density decreased from 0.883 by 14.2 percent.[8]

The lower bound for three dimensions needed no development, because it was known since Gauss's times that the best packing would have to have a density of at least 74.05 percent.

I must add a significant postscript to this chapter. In his second paper, Muder reported a momentous news item. A professor at Berkeley, Wu-Yi Hsiang, claimed to have proved Kepler's conjecture. In the bibliography, Muder cited a preprint by Hsiang from the previous year, "On the sphere packing problem and the proof of Kepler's conjecture." That was a real bombshell. While worthy mathematicians were fiddling with chain saws, kitchen knives, and scalpels, just to find better bounds, a professor on the West Coast had quietly made all these efforts superfluous. After 381 years Kepler's conjecture had finally been solved. Or had it? Muder had some doubts about the veracity of Hsiang's assertion. He expressed them in a caveat to his readers: "As of this writing, the status of [this claim] is unresolved."

---

[8] I limited the discussion to three dimensions in this chapter. However, some of the work that was discussed includes results that also apply to upper bounds in higher dimensions. I mentioned in chapter 4 that in 1905 Minkowski showed that the Riemann Zeta function provides an appropriate lower bound for higher dimensions.

CHAPTER 10

Right Angles for Round Spaces

In the late 1980s a new player appeared on the scene. It was Thomas Hales, an assistant professor of mathematics at the University of Michigan. Hales had impeccable academic credentials: B.A. and M.A. degrees from Stanford, a year at Cambridge University in England, and a Ph.D. from Princeton University. At Cambridge he sat, with distinction, for the Part III Tripos, an exam named after the three-legged stool on which the student traditionally sat when he was questioned. At Princeton he was awarded the Harold W. Dodds Honorific Fellowship. Teaching posts at Harvard and at the University of Chicago followed, interspersed with research appointments at the Institute of Advanced Studies at Princeton and the Centre National de Recherche in France. Since 1993 Hales has been on the mathematics faculty of the University of Michigan.

Hales heard about Kepler's conjecture for the first time in the autumn of 1982, in a course taught by John Conway at Cambridge. In 1988 he encountered the conjecture again. He taught an elementary undergraduate course in geometry at Harvard, and the textbook he used mentioned the problem. This is when he started thinking about it seriously. He started developing a strategy that he thought would lead to a proof. He had just got into high gear with his grand design when disaster struck. It was 1991, and he had been thinking about his scheme for the past three years, when a colleague informed him that a professor on the West Coast had found a proof of Kepler's conjecture. The information was not yet official since no paper to that effect had appeared in a respected journal, the seal of quality and the hallmark of scientific achievement, but various preprints were floating around the mathematics community. One of these preprints was mentioned by Doug Muder in his upper-bound paper. Finally, in 1993, the *International Journal of Mathematics* published "On the sphere packing problem and the proof of Kepler's conjecture," by Wu-Yi Hsiang of the University of California at Berkeley.

Predictably, for Hales the news went down like a lead balloon. Here he was, laboring away on a grand plan to prove a conjecture that nobody had been able to prove for the last 380 years when all of a sudden someone appears out of the blue and manages to snatch away victory. Couldn't Hsiang have waited a few years? Then priority would have been Hales's, and any other proof would appear under the also-rans. Of course, the genuine intellectual, the scientist solely concerned with furthering human knowledge, would not mind who gets priority, as long as a truth has been established. But human nature is different and such a scientist does not exist, at least these days, where notoriety, fame, and reputation count for just about everything. Hales did not want to believe it, but when the newspapers got wind of the story and made a big deal out of it, it dawned on him that the news might be true. What must have really got to him was the media hyperbole that greeted the feat. The 1992 Yearbook of the *Encyclopedia Britannica* called the feat "without doubt the mathematical event of 1991." *Science* wrote on March 1, 1991, that a professor at Berkeley had picked "the oldest, hardest unsolved problem. The second thing he did was solve it." And *Discover* gushed in its January 1992 issue that a math professor with a flexible mind "knocked off a proof" that it hailed as "one of the most remarkable achievements in the history of mathematics."

Despite Hales's disappointment, he still had reason to believe that not all his travails had been for naught. A false claim had been made before. No less a figure than Buckminster Fuller had previously claimed to have proven Kepler's conjecture, only to have done nothing of the sort. Bucky, as he was affectionately called by everyone, was the architect who lent his name to the Buckyballs, or Fullerenes, molecules that are shaped like his best-known invention, the geodesic dome.

The geodesic dome is a structure that maximizes space and uses building materials most efficiently. Mathematically speaking, a sphere maximizes volume while, at the same time, it minimizes surface.[1] But since it is difficult to build round walls and ceilings, Bucky designed a polyhedron that approximates a sphere. The geodesic dome uses only straight-line segments, but from afar it looks like a sphere. The advantages of the dome are too numerous to list. Suffice it to say that a U.S. infantry general remarked during the Korean War that the geodesic dome was the only real advance in mobile shelters since the tent had been invented a few millennia ago. Bucky's domes were the most useful gadget for survivalists since the Swiss army knife. There is only one small blemish: Bucky wasn't the first person to invent it. A certain Dr. Walter Bauersfeld was. He built a lightweight round structure made

---

[1] Remember Dido's problem from chapter 3?

up of polygons for the famous Zeiss Optical Works in Germany in 1922. The edifice was used as a planetarium on the roof of the factory, and people traveled from far away to catch a glimpse of it. Bucky was simply the first man to get a patent for the dome, exploit it commercially, and do a song and dance about it. And while we're at it, let's also mention that the Inuit built their igloos long before anybody had ever heard of either Dr. Bauersfeld or Buckminster Fuller. So much for American innovation.

But Bucky was not content to be an architect; he also considered himself a mathematician. One of his claims was that he had proved Kepler's conjecture. The proof was contained, so he asserted, in his magnum opus, the two-volume *Synergetics* (1975 and 1979). In fact, sphere packings are at the heart of *Synergetics*. References to it are scattered throughout the books. Bucky was very concerned with the densest packing of spheres, since it provided a prime example of stability and equilibrium. To Bucky, the architect, these two features were of overall importance. After all, who wants a house that falls down? Fuller called Kepler's arrangement of spheres, FCC and HCP, the "isotropic vector matrix," or "isomatrix" for short. In other places it is called a "vector equilibrium," or octet truss or cuboctahedron. All terms but the last one are of Bucky's invention. One should always be suspicious when an author tries to repackage old ideas by inventing new words. It is often a sign of forced originality when true innovation is lacking. But even though Bucky did not add anything uniquely new to Kepler's problem, he was nothing if not original.

In *Synergetics* Bucky launches into a long-winded discussion of how spheres arrange themselves in space if they are left to their own design, that is, to gravity. One sphere on its own serves no purpose because there is no one there to observe it. So a second sphere must be added to the universe, and from that moment on gravity comes into play. The two spheres are attracted to each other. Bucky maintains that two spheres, left to their own devices, will swirl around for a while until—under the influence of gravitation—they eventually knock into each other and arrange themselves in the form of a dumbbell without a handle. At this point a third sphere is added to the universe. After it has done its swirling around, it will dock onto the dumbbell right in the sweet spot, the crevice between the two first spheres. If it docks anywhere else, gravitation will immediately pull it towards the crevice. Only then is the ménage-à-trois stable. To top things off, add a fourth ball. After having done its share of whirling, swirling, and zooming around, it will automatically land in the nest created by the first three spheres. If it happens to land on one of the balls instead, gravity will soon make it roll into the nest. So there you have it: the four spheres have arranged themselves into a tetrahedron.

With five balls swirling around in the universe, another configuration is possible. First two pairs of spheres arrange themselves in two dumbbells and dock next to each other, thus forming a perfect square. This square is unstable, except that at precisely that moment a fifth ball drops into the nest in the middle. We now have a stable pyramid or, looked at in another way, half an octahedron. When the sixth ball comes swirling by, it will land in the nest on the opposite side, and a full octahedron will have formed.

Only when spheres are arranged in tetrahedra or octahedra are they at rest. Any other configuration is unstable. Bucky maintained that under the influence of gravity, balls would twitch, fidget, and wiggle until they settle in comfortable resting places, forming a tetrahedron or an octahedron.

Once we have reached this stage, everything else is easy. Release more and more spheres into the universe and let them swirl around. One after another they will land in the nests created by the previous spheres, thereby building up a larger and larger configuration. The result is a so-called cuboctahedron, or isomatrix, that grows thicker with every layer of balls. The first layer (after the "nucleus" in the middle) contains twelve balls, the next forty-two, the one after that ninety-two, and so on. The nth layer contains $10n^2 + 2$ spheres.

Looking closely at this cuboctahedron, one notes that not only the nucleus is surrounded by twelve spheres. Each sphere on its own is also surrounded by a dozen others. So there you have it: the FCC. In Bucky's words: "Being omnidirectionally equally interspaced from one another, this omni-intertriangulation produced the isotropic matrix of foci for omni-closest-packed sphere centers." Couldn't have put it better ourselves.

So what has Bucky done to prove Kepler's conjecture? In one short word: nothing. His "proof" was just a description of the face-centered cubic packing, albeit a very idiosyncratic one. Swirling balls—left to their own design and solely under the influence of gravitation—like to arrange themselves

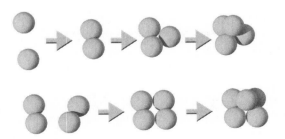

Bucky's balls, getting comfortable

neatly into a cuboctahedron. So? The configuration is stable, we'll admit to that. But does that imply that it is the densest packing? Not by any stretch of the imagination! Bucky's musings were certainly not a proof of Kepler's conjecture.

Wu-Yi Hsiang was the newfound darling of the scientific press. He received his B.A. from the National Taiwan University in 1959 and then completed two years of mandatory military service. Since there were no graduate schools in his native country, in the 1960s he went to Princeton where he got his Ph.D. in 1964. In 1968, during the heyday of the flower power revolution, he became professor at the University of California at Berkeley, the focal point of the Woodstock generation. San Francisco and Haight-Ashbury were just a stone's throw away, anti-Vietnam protests were the order of the day, and kids with flowers in their hair experimented with alternative lifestyles. Hsiang stayed at Berkeley for nearly thirty years; in 1997 he left the West Coast to answer a call from the Hong Kong University of Science and Technology. Wu-Yi's brother, Wu-Chung, is also a distinguished professor of mathematics at Princeton.

Hsiang's main research interests were transformation groups, global differential geometry, classical geometry, and celestial mechanics. Hsiang is an inspiring educator who took his teaching duties at Berkeley very seriously. One semester he announced a course on classical geometry. In order to get his students interested in the subject, he sought an interesting application of the theoretical concepts and hit upon Kepler's conjecture. Once he started thinking about the problem, it would not let him go.

He began to fiddle around with the conjecture, and after months of wrestling with the problem he was satisfied that he had, in fact, succeeded in proving it. Not only that, but at the same time he also achieved a proof for the dodecahedral conjecture, the remnant of Fejes-Tóth's failed attempt to prove Kepler's conjecture. He had killed two birds with one stroke—or so he believed.

How had he done it? Earlier attempts to prove Kepler's conjecture, for example Fejes-Tóth's foray in the 1940s, compared wasted space in different packing arrangements, that is, the combined volume of all the gaps that were left between the spheres in that arrangement. Then mathematicians would vary the packing arrangement a bit and check whether the wasted space became larger or smaller.

But that was easier said than done. The accepted method to compute wasted space was to subdivide the whole space into many-sided cells, the V-cells. Each of them contains one sphere and some wasted space. The shape of the V-cells is determined according to which packing arrangement is used and, accordingly, the wasted space also varies. Loose packings imply large cells with lots of wasted space, dense packings use small cells with less space wasted. The key to the proof—so it was thought—was to compute the wasted space in each packing arrangement.

The tools were provided by convex polyhedral theory. In this theory, cells are divided into simpler shapes (polyhedra) and then the wasted space is calculated in each of these polyhedra. But mathematicians who tried their hand at it soon found out that this approach led nowhere. Hsiang decided that something different was needed.

Up to that time, mathematicians had used Cartesian coordinates in their attempts to prove Kepler's conjecture. Cartesian coordinates were invented by, and are named after, René Descartes, the seventeenth-century philosopher and mathematician. They work as follows: in a two-dimensional plane, the position of a point can be characterized by stating its distance from some origin in each of two directions. It is like a pirate's map, which indicates that Captain Crook's hidden fortune is buried 60 meters to the east and 80 meters to the north of the Big Oak Tree. These indications leave no doubt as to the whereabouts of the hidden fortune, and if the chest hasn't been dug up before then the treasure hunter would have no trouble finding it.[2] So 60 meters east and 80 meters north are the Cartesian coordinates of the treasure chest.

The Cartesian coordinate system, which is the basis of Euclidean geometry, is inherently rectilinear: right angles and straight lines, horizontal and vertical, form its basis. The problem with the Cartesian system for Kepler's conjecture is that spheres, and the wasted space hugging them, is inherently curved. In Hsiang's words: "Spheres—the most symmetric bodies—are perfectly round in shape, while the whole space is basically rectilinear in nature. Since roundness and rectilinearity clearly cannot fit together very well, there always exists a considerable amount of unfilled interstices for any sphere packings."[3] Hsiang decided to try his luck with something different: spherical coordinates.

Spherical coordinates also need an origin, but instead of specifying two directions the new system makes do with just one, for example, the north. The position of a point in the plane is then defined by again specifying two numbers: the direction and the distance from the origin. We could say that a point lies 100 meters northeast of the origin.[4] Hsiang thought that spherical

[2] The map should also indicate how deep the chest is buried. The three numbers together define the chest's location in *three* dimensions.
[3] This quote is edited slightly.
[4] Mathematicians rarely use the navigator's compass rose, and usually say "45 degrees" instead of northeast. North is defined as zero degrees, west as 90°, south as 180°, and east as 270°.

geometry would be the fundamental tool in solving Kepler's conjecture, and that it would be better able to deal with the problem.

One of the advantages of spherical coordinates can be seen in the following example. A treasure hunter using a Cartesian map and starting from the Big Oak Tree would walk 60 meters to the right, make a left turn, and walk another 80 meters to reach the point where the chest is buried. A treasure hunter using a spherical map, on the other hand, would walk 100 meters in the 31° direction. Both would get to the same point, but the spherical walker would already be busy digging for the chest by the time the Cartesian walker arrived.

Some people, for example Arthur L. Loeb, the chemist/physicist/ choreographer-turned-professor of design science from the Department of Visual and Environmental Studies at Harvard University, argue that spherical coordinates are more natural than Cartesian coordinates since 90° angles usually do not appear in nature. Like Buckminster Fuller, Loeb maintains that right angles, squares, and cubes are simply artifacts of architects, mathematicians, and modern artists. In his freshman seminars on Structure in Science and Art, he encourages students to explore space by emphasizing natural structure. He asks them to avoid right angles, which do not prevail in the art of people in closer touch with nature. This, by the way, is why the huts of African tribes and the Inuit's igloos are usually round.[5]

Let us return to Kepler's conjecture. A sphere packing consists of balls that either touch each other, or that are close but do not quite touch. The first thing Hsiang did was to delineate a tight polyhedron around each sphere and

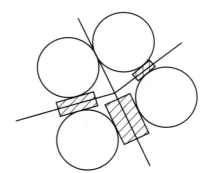

Polyhedra and slabs

---

[5] The absence of right angles and straight lines in art and architecture has a drawback, however. Some members of African tribes are said to have difficulty with depth perception, which may be due to the lack of right angles in their day-to-day experience.

check how much of its volume stays empty. If two spheres touch, the face of the polyhedron contains the point of contact. If two spheres are close but do not touch, the face of the polyhedron runs through the middle of the empty space that is left in between. We recognize the familiar V-cells. In the first case the ball reaches all the way to the face of the polyhedron. In the second case, some space is contained between the sphere and the polyhedron's face. Hsiang called this space a *slab*.

Hsiang's approach consisted in classifying the interstices—the nooks and crannies nestled between the spheres—into two categories: the *peripheral* interstices, which consist of the slabs, and the *core* interstices, which contain everything else. Since the slabs are box-like shapes, with angles and straight edges, the volume of the peripheral part can be computed with Cartesian coordinates. That proved to be the easy part. The core interstices, which contain the wasted space that arches around the spheres, are curved. The determination of their volume requires more sophisticated tools, and this is where spherical geometry came in handy. First Hsiang subdivided the core part into subpolyhedra, using the faces of the V-cell as bases. Then he derived formulas for their volumes. The lower limit of the slabs' volumes provides the upper limit for the density.

Next, he examined how many spheres could approach a central ball. On the one hand, space will be overcrowded if the number of balls is thirteen or more, while, on the other hand, a considerable amount of gaps will be left between the balls if their number is twelve or less. Hsiang sought to

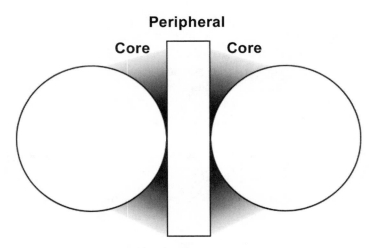

Peripheral and core interstices

understand *overcrowdedness* in the former, while studying the geometry of *non-touchingness* in the latter.[6]

Up to this point Hsiang's proof only considered local density. By *local,* Hsiang meant a configuration that consisted of the central sphere, its neighbors, and the neighbors' neighbors. This he called the "double layer local packing." The double layer local packing of Kepler's proposed configuration, for example, consisted of the central sphere, the twelve immediate neighbors, and the forty-four neighbors of the twelve immediate neighbors (a total of fifty-seven spheres).[7] The double layer approach allowed Hsiang to compute local density. But how can these computations be extended to the third, fourth, and further layers? In simply adding all wasted spaces, without taking account of the fact that some of them are common to more than one neighborhood, one would be guilty of double counting, and would overstate the wasted space.

So Hsiang was faced with a problem: How would the calculations, which only examine local neighborhoods, be extended to infinite space? Eventually he saw a way out. While local density cannot be a proxy for global density, the average of the local densities can. Hsiang then showed that "the average local densities cannot be greater than 74.05%." To prove that no arrangement can be tighter, he examined different types of packings, one by one. First he looked at configurations that can be formed by a central ball, surrounded by at most twelve close neighbors. Then he examined configurations with thirteen or more close neighbors.[8] In each of these cases he proved that his contention was true. From this he drew the general conclusion, that no packing arrangements exist with local densities greater than 74.05 percent. Since this is exactly the density of the face-centered cubic, this statement implies that Kepler's packing is the tightest possible.

Hsiang was elated. After nearly four centuries, he was the first man to provide the proof of one of the oldest unsolved problems in mathematics. He traveled the continents, attending workshops and giving lectures on his proof. Everybody was very excited. But gradually doubts arose: Hsiang glossed over the nitty-gritty of his proof, and his presentations did not contain sufficient detail, so nobody could quite follow them. Nevertheless, his talks were always vague enough to be plausible. Finally, in the summer of 1990, Hsiang sent out preprints of his proof to colleagues around the world.

---

[6] Sometimes it is difficult to find an expression for an abstract mathematical concept. However, mathematicians are never at a lack of esoteric words. Hsiang even has a plural for *nontouchingness,* which is, of course, *nontouchingnesses.*

[7] We recognize the first two layers of Bucky's cuboctahedron.

[8] Two spheres are defined by Hsiang as close neighbors if their centers do not lie more than 2.18 radii apart. They do not necessarily have to touch each other.

The distribution of working papers and preprints represents a convenient way to share not-quite-finished work with colleagues in order to get their feedback and, equally important, to establish priority. Sometimes preprints contain work that has been accepted for publication, but—given certain journals' backlog—may not see the light of day for many more months. At other times working papers consist of unfinished work, usually marked "not for quotation," that is sent to everyone on the author's mailing list in the hope of receiving some feedback. Hsiang made ample use of the latter method and was soon bombarded with requests for clarifications on points that had been left obscure. He soon realized, or was told by colleagues, that the preprint contained shortcomings, and in the following months Hsiang issued revisions of the preprints and revisions of the revisions.

But his colleagues were not happy with Hsiang's stopgap manner and soon people began to ask questions. The professor got exceedingly frustrated with the persistent inquiries by mathematicians who, in his mind, didn't quite understand as much about geometry as he did. Finally he decided that enough was enough and that his result should be officially presented to the world of mathematicians. On November 17, 1992, he submitted the hundred-page proof to the *International Journal of Mathematics*.

Publication in a respectable journal is generally considered the seal of approval, and Hsiang hoped that this would stop the muttering. Nevertheless, *Science* cautioned that his work must still be checked by the mathematics community, and that the 400-year-old quest would only be over "if his proof holds up under the [mathematicians'] gaze." And *Discover* hedged its characterization of the proof as one of the most remarkable achievements in the history of mathematics with the caution, "if it holds up."

For Hsiang it was very convenient that the *International Journal of Mathematics* was edited by colleagues from the Berkeley mathematics department. His acquaintance with the editors would undoubtedly speed up the refereeing and publishing process.

As Hsiang had hoped, his paper got a quick and sympathetic reading. But the editors did not want to give the impression that they had accepted a colleague's paper without checking it for rigor and completeness, and asked for a revision. On March 9, 1993, Hsiang submitted a revised version and it was duly accepted.

Allow me an aside about the refereeing process in professional journals. Once a manuscript has landed on an editor's desk, it takes some time to decide on the referees, and then the paper is sent out to them. If the chosen referees are not on sabbatical somewhere, a fact of which the editor may become aware only after not having received any response for a few months, the refereeing process can begin. But this is just the beginning. Any self-respecting professor considers himself far too busy with his own research, conferences, and administrative duties (let alone teaching) to find the time to do the refereeing. So he or she first sits on the paper for a couple of months. Then the paper may undergo a preliminary leaf-through. Finally the referee finds the time to read it thoroughly and write a report together with a recommendation to the editor. By that time, five, six, ten, or even more months may have passed. The editor then may take another month or two to make a decision. The net result is that a year's wait is nothing unusual. And the whole process may start over if a revision is needed.

The reader may note that the time between the submissions of Hsiang's original paper and the revised version was less than sixteen weeks! Obviously four months is an inordinately short time for one or more referees to check a ground-breaking, extremely bulky paper and write a report; for the editor to make an initial decision and send it to the author; for the author to revise the paper along the suggested lines; and finally for the editor to check and accept the revised version.

Doubts about the seriousness of the refereeing process seemed more than called for. Doug Muder, the upper bound man, was prompted to comment that the *Journal*'s refereeing process had obviously not worked as it was supposed to. "Hsiang's paper was not adequately refereed, if it was refereed at all. The fact that the *Journal* is edited by Hsiang's Berkeley colleagues lends an air of cronyism to the story. It seems clear that Hsiang chose the *International Journal* because it was edited by his friends."

The fact that the refereeing process was flawed does not necessarily mean that the paper was wrong. It could still contain a valid proof of Kepler's conjecture. But it didn't. And it did not take long for the fallout to hit Berkeley. The first experts soon expressed reservations about Hsiang's proof, and the full extent of its shortcomings became apparent soon after. Gábor Fejes-Tóth of the Hungarian Academy of Sciences, son of Laszlo and himself a mathematician of note, made an assessment for *Mathematical Reviews,* a journal that reports on the correctness and importance of articles from the hundreds of mathematics journals around the world. He wrote, "many of the key statements have no acceptable proof," and continued, "this cannot be considered a proof. The problem is still open." He concluded with a resounding denunciation. "If I am asked if the paper fulfills what it promises in its title, namely a proof of Kepler's conjecture, my answer is: no." In 1997 Károly Bezdek, a colleague of Hsiang's from Eötvös University in Budapest, wrote that "Kepler's conjecture . . . [is] still unproven. [Hsiang's] work is far from being complete and correct in all details." Thomas Hales agreed: "This problem is still unsolved. I haven't solved it. Hsiang hasn't solved it. Nobody else has solved it as far as I know."

Bezdek spent more than a year working with Hsiang in an attempt to fill in the gaps. In the end he gave up. The proof was simply wrong. Bezdek submitted a paper to the *Journal* detailing a counterexample to one of Hsiang's central claims. This time the editors took their time. They were in no hurry to publish a counterexample to the *Journal*'s most publicized paper in many years. Only after some delay was Bezdek's paper accepted and published in the *International Journal of Mathematics,* in 1997.

By and by, the deficiencies of Hsiang's proof became apparent and were roundly criticized in workshops, at conferences, during afternoon teas in mathematical departments around the world, and in private conversations by almost everybody who was anybody. Soon the mathematical community was nearly unanimous—Hsiang's proof was not what it claimed to be. A letter from Doug Muder to Barry Ciapra, the writer for *Science,* put it succinctly: "1. Hsiang's paper . . . is not a proof of the Kepler conjecture. At best it is a sketch (a 100 page sketch!) of how such a proof might go. 2. Even as a sketch the paper is inadequate, since counter-examples have been found to several of its steps. 3. Hsiang's related claim to have proved the Dodecahedron conjecture . . . is equally baseless. 4. Work on the Kepler conjecture and the Dodecahedron conjecture should continue as if Hsiang's paper had never existed." In the ivory towers of academia vicious battles are not uncommon. But Muder's comments used strong language indeed.[9]

Given the stringent laws of mathematics, we would have expected the members of the profession to agree on whether a statement or a collection of statements from which a proof is deduced is true or not. After all, mathematics—unlike the justice system—needs no interpretation of facts. It is an exact science, and feelings or prejudices play no role in deciding whether a proof is true or false. This is also not medical research, wherein patient records could be falsified and scientific results doctored. Everything is out in the open and everybody versed in the language of mathematics should be able to see for himself or herself whether a proof is correct and a statement is true.

But mathematics is not quite value-free, after all, and interpretation may still play a role, if only for a while. After sufficient time has passed for the experts to scrutinize every single step of a proposed proof, the community of mathematicians decides, by majority if not by consensus, whether a paper should be considered correct. In the case of Hsiang vs. Kepler's conjecture,

---

[9] Henry Kissinger, the former Harvard professor and secretary of state, was once asked why departmental fights are so violent, why back stabbing is so common among academic colleagues. His answer was short and to the point: "Because the stakes are so small."

the jury soon came back with a verdict: guilty! Hales could breathe a sigh of relief. He was still ahead.

What did the jury find? John Conway, Tom Hales, Doug Muder, and Neil Sloane announced in a letter to the *Mathematical Intelligencer* that they had objections to Hsiang's proof, and that Hales would describe the alleged holes in a forthcoming article. The *Mathematical Intelligencer* is a widely read publication with information, news, and historical tidbits about mathematics and mathematicians. It is not a peer-reviewed journal, and articles that appear there may contain opinion pieces that do not reflect the beliefs of anyone but the authors.

Before setting out on his debunking article, Hales wrote letters to Hsiang asking for clarifications on some subtle points. He didn't get clear answers. Worse still, Hsiang resorted to insults. "Your letter made me realize that I cannot assume that the average mathematician knows too much about elementary spherical geometry," he wrote. To claim that the Stanford-, Cambridge-, and Princeton-educated Hales was an average mathematician who knew little about elementary geometry was a bit low. Hales was quick to counter, and reprimanded Hsiang for his habit of giving proofs by handwaving: "Students who resort to such tactics jeopardize their grade-point average. For a professional, it is hardly imaginable." This did not sit well with Hsiang. In a rejoinder, published in a subsequent issue of the *Mathematical Intelligencer,* Hsiang wrote that "a fake counterexample . . . [is] manufactured, . . . easily provable statements are tortured into fallacious statements," and that Hales & Co. sometimes take refuge in "outright misrepresentation . . . to explain away their own misunderstandings." That is it, then: Hsiang regards Hales as rather less than an average mathematician; Hales considers Hsiang as being less capable than an average student. Not often did the venerable pages of the *Mathematical Intelligencer* carry such a barrage of verbal fire.

Mathematicians considered Hsiang's rejoinder inadequate; it could not cover up the proof's deficiencies. But Hales was not willing to waste further time poking additional holes into his colleagues' paper. Debunking Hsiang's rejoinder would continue a never-ending cycle that he simply did not have time for. As things stood, he would not be able to convince Hsiang anyway, and so he just dropped the matter.

What was the problem? In the *New Scientist,* in 1992, Ian Stewart of Warwick University cites two problematic statements that Hsiang had made in his preprints. Both of them are hair-raising. In one example, Hsiang allegedly stated that the surface of one triangle is larger than the surface of another triangle if its edges are longer. While this is true for equilateral triangles, a picture easily convinces the reader that it is clearly incorrect as a general statement about all triangles. All edges of Triangle A are longer than the edges of Triangle B. Nevertheless, the surface of Triangle B is greater.

A counterexample suffices to disprove a statement, so one incorrect link in a chain of arguments invalidates any mathematical paper. The situation is similar to one in a court of law as was brought home by the lawyers in the murder trial of O. J. Simpson. Even though much of the evidence pointed to Simpson's guilt, doubt on just one item in the chain of evidence sufficed to allow the jury to declare the former football star innocent of the murder of his ex-wife. A mathematical proof is no different.

But there is more. According to Hales and Stewart, Hsiang claimed that if a number of objects cannot be contained within a certain region, then a smaller region can not contain them either. Again this is not true. A glance at the picture shows that two circles cannot be placed into a square of surface 9, while a rectangle of surface 8 is able to contain them. Did Hsiang, an accomplished mathematician by all accounts, really make such foolish errors? In his rejoinder to Hales's broadside, Hsiang denied he used such defective arguments in his proof. He claimed that Hales, Stewart, and all other critics attributed fallacious statements to him. In the published version of the proof these errors, in fact, no longer appear. But in the preprint they did, and even though Hsiang eventually managed to patch up some of the errors, his credibility in the mathematics community was badly tarnished.

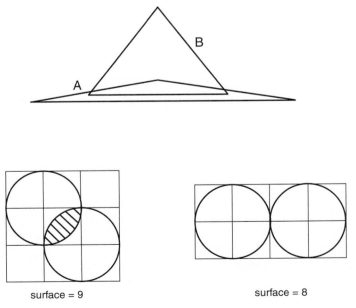

surface = 9

surface = 8

Hsiang's errors

Should a mathematical paper depend on the reputation of the author instead of being judged on its merit? It should not, and let's not belittle the published paper because of the deficient preprints. But as turns out, the published version of the proof also contained numerous errors, fallacies, and gaps. More than once Hsiang used a dubious trick, which Hales disparagingly called "critical case analysis." A general statement was reduced to a few critical cases. Hsiang proved the statement for the critical cases and then bombastically—and without proof—claimed that the statement is correct for *all* cases. This is quite inadmissible. Furthermore, the paper is replete with phrases like "it is quite easy to show," "it suffices to consider," or "the same method will imply the general case." Well, it usually was not easy to show, it did not suffice to consider, and the same method did not imply the general case.

At some point in his proof, Hsiang derived a lower bound for something he called the "uniform buckling height." The numerical value of this bound turned out to be 0.0316. The estimate Hsiang needed, however, was 0.0250 for *all*, not just for the *uniform,* buckling heights. No problem: "Since 0.0316 is more than 25% larger than 0.0250, [this] also implies the lower bound of 0.0250 . . . in the non-uniform case." That really takes the biscuit! Just because one object is much larger than a certain numerical value does not mean that another object is also larger than this value. Hales compared Hsiang's assertion to the following ridiculous statement: "If the foothills, which we see, are no more than 1,000 feet, then the mountain peak, hidden by the clouds, cannot be more than 1,250 feet."

Hsiang himself remained unconvinced. He was well aware of the counterexamples, and of the fact that his claims were not believed by the experts in the field. Nevertheless, he considered the whole hullabaloo about his paper to be no more than a discussion about how much detail must be included in the published proof. One is reminded a bit of a childish game of one-upmanship. If I gloss over a few steps in the proof, and you get lost, then I must be much smarter than you are. If I state that the truth of a statement "can easily be seen," and you don't see it, then you obviously do not measure up to my high standards. Hsiang glossed over too many steps, and declared too often that part of an argument can easily be seen. Paraphrasing Abraham Lincoln: one can fool some mathematicians some of the time, but one can't fool all mathematicians all of the time.

According to John Conway, nobody who read Hsiang's proof has any doubts about its validity: the paper is nonsense. With this he did not mean Buckminster Fuller's kind of nonsense, but nonsense all the same. Hales was a bit more diplomatic. He was prepared to admit that despite its shortcomings there was much of value in Hsiang's paper, but that he destroyed the credibility of his own work in trying to present experimental hypotheses as if they were facts. Mathematicians, Hales claimed, can easily spot the difference between handwaving and proof. Eventually, Hsiang's proof was declared dead, and quietly carried to its grave. It would have been nice to have had an elegant, analytical proof that used nothing more than well-known tools from spherical geometry, vector algebra, and calculus. But for the time being, it was not to be.

In spite of all the flak, Hsiang remains to this day deeply convinced of his proof's correctness. And one cannot simply discount his generally brilliant geometric ideas and insights. Colleagues recount that Hsiang is sometimes able to see more in a simple sketch than others are able to discover in a whole picture book. Everything he writes inspires those around him. Unfortunately, much of it is incomplete and not correct in all details. Furthermore, Hsiang seems genuinely unable to understand why his colleagues are often dissatisfied. His mind lives in another world and he is often not responsive to doubt and criticism. In science this can be disastrous.

Maybe Hsiang's proof will be cleaned up one day. Initially, even Conway seemed to believe it could. Then, on seeing Hsiang's intransigence, everyone gave up. Most troubling is the fact that while some of the building blocks that make up Hsiang's "proof" may eventually be proven, others are factually wrong. Nevertheless, Bezdek felt that "a combination of Fejes-Tóth's and Hsiang's strategies with some other methods might work." But, he added, "one needs further combinatorial and analytic ideas in order to have a proof." What was already clear, however, was that it would take more than just a little cleaning up.

Recently, Hsiang made another attempt to rehabilitate his proof. He had come to realize that at the very least his paper was lacking in detail. To fill in the holes he published a book in December 2001 with World Scientific entitled *Least Action Principle of Crystal Formation of Dense Packing Type and Kepler's Conjecture*.[10] The publisher's advertisement heaped praise on it: "This important book provides a self-contained proof of [Kepler's conjecture], using vector algebra and spherical geometry as the main techniques . . . in the tradition of classical geometry." Whether the work will live up to the publisher's expectations remains to be seen. Close scrutiny by a multitude of mathematicians will be required to ascertain whether the gaps have been filled. One noteworthy detail: the work of Tom Hales is not mentioned once.

---

[10] Hsiang feels that the optimal *finite* packing (as opposed to Kepler's infinite packing) is actually more useful for the understanding of nature.

CHAPTER 11

Wobbly Balls and Hybrid Stars

The Hungarian professor and mathematician Laszlo Fejes-Tóth solved the packing problem in two dimensions in 1940 and then turned his attention to the three-dimensional problem. There, however, his luck ran out. As described in chapter 9, his attempt to establish a new upper bound ended in total failure. Subsequently, his futile endeavor went into history as the dodecahedral conjecture.

Fejes-Tóth believed that nature has a tendency to organize its building blocks into regular structures. Nature's inclination towards efficiency (for example, by minimizing a volume or maximizing a density) should automatically bring about regularity: maximize density and what you find is a regular lattice. Put another way, regularity in nature is usually brought about by some efficiency principle: scratch a regular structure, and an efficiency problem is hidden somewhere. Thus order arises out of chaos. Fejes-Tóth's deeply held beliefs anticipated the notions of chaos and self-organization dozens of years before they became fashionable. A case in point would be Kepler's sphere arrangement. Taken on their own, the cells that make up Kepler's arrangement are not the densest at all. But they tile space in a regular fashion, on a lattice, and thus should represent the densest global arrangement. Fejes-Tóth's ideas are in general thought-provoking, but without a formal proof they sound much like the unsupported speculations of his contemporary, Buckminster Fuller.

Undaunted by his failure with the dodecahedral conjecture, Fejes-Tóth returned to sphere packings time and time again. In 1953 he wrote a book in German entitled *Lagerungen in der Ebene, auf der Kugel und im Raum* (Packings in the plane, on the sphere and in space). It summarized everything that was known about the subject at that time. The book's last chapter was devoted to packings in three-dimensional space. Fejes-Tóth could not stop himself from repeating the claim he had made ten years earlier, that no sphere packing could be denser than 75.46 percent. He knew, of course, that the evidence he had offered in support of the dodecahedral conjecture was lacking. But this did not keep him from presenting the "not quite exact, but from a certain point of view nevertheless rather satisfying proof." It is not clear which point of view Fejes-Tóth had in mind, but obviously no such thing as a "not quite exact proof" exists.

Fejes-Tóth then suggested a two-stage proof for Kepler's conjecture. The first stage would consist of dividing space into Voronoi cells. (Recall that a sphere packing's V-cells are polyhedra whose walls run exactly halfway between two neighboring spheres.) The next stage would consist of the search for cells with the smallest volume. The regular dodecahedron would have been a good candidate. It is small all right. But as a V-cell it is quite useless, since it cannot be built into a gap-free packing of the entire space. So the question arises: What is the purpose of examining a V-cell? One may happen upon a good candidate, only to find out that it cannot tile space. Then, in order to fill the invariable gaps, neighboring V-cells must be of different shapes. But when gluing differently shaped cells together a small V-cell is usually surrounded by large, loosely fitting V-cells. The unused volume of the loose neighbors wipes out the advantage of the tight fit in the center. This is what happens with the regular dodecahedron.

When Fejes-Tóth became aware of this, he realized that he had been following the wrong approach. To examine V-cells on their own is too short-sighted. The central V-cell must be considered together with its neighbors, and the neighbors' neighbors, and the neighbors' neighbors' neighbors. The key to Kepler's conjecture was the examination of a whole cluster of spheres simultaneously. Then, instead of considering the volumes of single cells, the *average* volume of the cells in the cluster must be computed.

That was the key, but where was the lock? Since there are infinitely many neighbors, how can the plan amount to anything useful? While pondering this question, Fejes-Tóth had a crucial insight. It should suffice to analyze finite clusters of balls and to find an upper bound for their packing density. Then there would be no need to compute the densities of an infinite number of spheres, which is what his predecessors had tried to do. Specifically, he proposed the following procedure: consider a collection of balls of unit radius that lie within a larger globe. Let us say that the maximum number of balls that fit into the globe is N.[1] The rest of the procedure is as simple as counting one, two, three. One, form V-cells around each of the balls. Two, compute the V-cells' volumes. Three, identify the densest cluster composed of at most N balls.

---

[1] An upper limit for the number of balls of radius 1 that fit into a globe of radius R is R^3.

You could think of the cluster as a bunch of grapes. The first task consists of gift-wrapping each grape in its own little box. Immediately a difficulty arises: the grapes in the outer layer are not completely encircled by neighbors. Consequently, the outermost grapes are al fresco, so to speak; some gift boxes have no exterior walls. Fejes-Tóth's way around this problem was to envelop the cluster with an additional layer of balls, which would be like placing the bunch of grapes into a partially filled container and then covering it with more grapes. Once a sufficient number of neighbors surround the central bunch, the outside walls of the gift boxes are identified.

The second task is easy. It involves no more than computing the average volume of the bunch's gift boxes. The third, and final task consists of choosing from among all possible bunches with N grapes or less the one whose gift boxes have the smallest average volume. Having completed these three tasks, the whole, infinitely large space may be filled up, using any clusters with up to N grapes. The average density of this *global* packing can never rise above the density of the best bunch. In this manner an upper bound for the packing density would be found.

If this bound turns out to be equal to 74.05 percent, Kepler's conjecture would be proven. The FCC and the HCP have exactly that density, and it would have been shown that no packing exists with a density greater than that. But what if the bound that has just been found is lower than 74.05 percent? Or if it is higher? Where would that leave us?

The first alternative can be dismissed out of hand: you will never find an upper bound that is lower than 74.05 percent because we know for sure that Kepler's packing achieves this density. But a higher bound *is* a possibility. (The dodecahedral conjecture, with a density of 75.46 percent, points ominously in this direction.) In this case a global packing that is better than Kepler's arrangement may possibly exist; it would just not have been found yet. Therefore, the search must go on. Either the mysterious packing with the higher density is found, or the upper bound must somehow be lowered.

The chances of finding a packing that is denser than Kepler's arrangement are extremely slim, so the latter way is the way to proceed. The only hope of tightening the upper bound is by increasing the size of the clusters. Let us, therefore, expand our horizon beyond the immediate neighbors of the middle ball, and include the neighbors' neighbors and, if need be, the neighbors' neighbors' neighbors. (Of course, one additional layer is always required in order to determine the gift boxes of the outermost balls.) After each new layer we seek the cluster with the highest average density. If this density turns out to be 74.05 percent, we're done; otherwise we increase the bunch once more and try again. Et cetera, et cetera, ad nauseam. The hope is that the upper bound will drop further and further, until it eventually hits 74.05 percent.

But where do we begin? Fejes-Tóth suggested starting with clusters of balls whose centers lie within a distance of 2.0534 of each other. Whence this magic number? Fejes-Tóth claimed that this is the closest distance at which thirteen balls can approach a central ball. At any distance less than that, no more than twelve neighbors can squeeze in.

Why is that? We know that within a distance of 2.0 from the center, the ball in the middle can have no more than twelve neighbors. Newton already said so.[2] But if the permissible distance is increased to a distance of 2.0534 anything could happen. Maybe thirteen balls could get that close. Or fourteen. After all, it took two and a half centuries to prove Newton's kissing problem, so let's not take anything for granted. In Fejes-Tóth's exposition we look in vain for a proof of the claim. Rather, the professor resorts to the time-tested method of handwaving. The paragraphs following his initial statement are replete with phrases such as "it may be assumed," "most probably," "it accords with experience," "it agrees with intuition," "presumably," and "there is no doubt that."

Who would dare argue with such intimidating proclamations? But coercion is no substitute for proof and Fejes-Tóth seems to have sensed that he was not being very convincing. So he delved into a long-winded explanation. "It's not really worth while to actually prove the inequality because it carries little interest." Sound persuasive? "The inequality should be considered a well tested empirical fact." Nice try, but no, thanks. "Anyway, the inequality is only used as a rough estimate which will only serve to exclude special cases." Fejes-Tóth's previous rough estimate of 1943, which was meant to exclude special cases, turned out to be unprovable. "A weaker inequality also suffices." It may also suffice, but it is also not proven.

Having made this passionate though not very compelling point, Fejes-Tóth pressed on without looking back. Limiting his investigations to configurations of spheres whose centers lie within a distance of 2.0534, he made two crucial assumptions. If these assumptions were true, he wrote, Kepler's conjecture would be proved. This time it was okay to make assumptions since they were used for illustrative purposes only. The first assumption was that the densest packing would not contain more than twelve neighbors—in addition to the central ball. Mindful of the above arguments, we can admit that such a statement may not be totally unreasonable. At least it was explicitly used as an assumption and not as a proven fact.

The second assumption was that in the densest packing the balls of the cluster should not be able to wobble. This assumption also seems reasonable, since a wobbly ball means that there is too much room left in the V-cell.

---

[2] At a distance of 2.0 the surrounding balls must kiss the central ball. And no more than twelve balls can do that.

Unfortunately, reasonableness is no substitute for a proof either. True to form, Fejes-Tóth again offered none. Instead he resorted to phrases like "[this cluster] appears to be the most advantageous" and "therefore one may assume" that this cluster achieves the minimum density. If the two assumptions were true, Kepler's conjecture would be proven: twelve balls arranged around a central ball in a rigid manner would be the densest cluster, and this cluster could be extended to infinite space. This is what Kepler had claimed. But since Fejes-Tóth's demonstration was based on unsubstantiated assumptions, he had not proved anything.

But that wasn't the point. The point was that for the first time someone had suggested that Kepler's problem could be reduced to a finite number of variables. All one had to do was minimize the average volume of the cells in the cluster. Fejes-Tóth had a hunch that the number of balls that would eventually have to be considered simultaneously would be about fifty: one ball in the center, a dozen neighbors, and close to forty that surround the twelve neighbors. These forty balls determine the gift boxes of the twelve neighbors. But the number could be much higher. V-cells have been discovered that have forty-four neighbors.[3]

Let us assume that Fejes-Tóth was right and that we must consider no more than about fifty balls. The task would be to find, among all possible configurations, the one with maximum average density. Since the positions of fifty balls in three-dimensional space are determined by their 150 coordinates, the objective would be to maximize a function of 150 variables. Fejes-Tóth

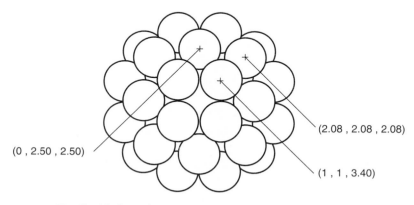

(2.08 , 2.08 , 2.08)

(0 , 2.50 , 2.50)

(1 , 1 , 3.40)

V-cell with forty-four neighbors (coordinates in parantheses)

[3] And it is not even certain that this number is the maximum number of neighbors possible. The best upper limit currently known is forty-nine.

summarized his thoughts with the following words: "We have indicated a concrete program for the solution of the sphere packing problem. Thus we have come a step closer to settling the problem."

But wait a minute. Maximizing a function of 150 variables is no trivial matter. The classical way to maximize a function consists of differentiating it with respect to each of the 150 variables, setting each expression equal to zero, and then solving the system of 150 nonlinear equations simultaneously in 150 variables. It would have been a daunting task. Fejes-Tóth fully realized this, but he remained confident. "Even though an exact treatment of this minimizing problem seems to be rather difficult, it cannot be considered hopeless." Rather difficult? The treatment may not be quite hopeless, but it is very difficult. Fejes-Tóth had taken a giant step forward by pulling the problem down from an infinite number of variables to only 150. But a solution was still light-years away.

Does this place us back at square one? Not quite. In 1965 Fejes-Tóth wrote another book, entitled *Reguläre Figuren* (Regular figures). It is a beautiful work about ornaments and mathematics. Stashed away in an envelope behind the front cover are stereographic images (if they haven't been stolen by a previous library user). They can be viewed in three dimensions with a special pair of eyeglasses that are stashed away in an envelope under the back cover. Toward the end, Fejes-Tóth returned to Kepler's conjecture. Most of what he had written on the subject twelve years earlier was reproduced here, nearly verbatim. But then Fejes-Tóth added one significant sentence, which paved the way for future attempts to prove Kepler's conjecture. "Mindful of the rapid development of our computers, it is imaginable that the minimum [of the volume function] may be approximated with great exactitude."

At that time, computers were still in their infancy. ENIAC (Electronic Numerical Integrator and Computer), the machine that ushered in the information age, had been built at the University of Pennsylvania's Moore School of Electrical Engineering less than twenty years earlier. The ENIAC was 30 meters long and 3 meters high. Its nineteen thousand vacuum tubes, fifteen hundred relays, and hundreds of thousands of resistors, capacitors, and inductors filled an entire room. It weighed 30 tons, and consumed 200 kilowatts of power. ENIAC generated so much heat that it had to be placed in a room with an air cooling system. Its speed, however, was truly awesome: ballistic trajectories that had taken a hand calculator twenty hours to calculate (and Lagrange's students twenty days) were computed by the ENIAC in just thirty seconds.

But speed is a relative term, and a pace that was considered awesome in 1946, and even in 1965, would be regarded as pedestrian today. Forget about processing speed measured in megahertz and information storage measured in gigabytes. In the mid-1960s, electronic computing was still an excruciatingly slow process. ENIAC could complete 5 KIPS (5,000 instructions per second). The 8088 chip introduced by Intel in 1979 was fifty times faster, at 250 KIPS. The 80486 processor, installed on the most modern PCs of 1989, was able to process an impressive 20 MIPS (20 million instructions per second). Four years later the Pentium chip processed 60 MIPS, and in 1995 the Pentium Pro was whizzing away at 200 MIPS. In early 1999 IBM launched the S/390 G6 Mainframe Server, a machine with a processing speed of 1.6 BIPS (1.6 billion instructions per second), and at the end of 2000 Hitachi announced a server that approached 3 BIPS.

In the mid-1960s, however, when Fejes-Tóth wrote his prophetic words, most computers still performed at the KIPS level. Even the setup for a computer run was agonizingly slow. It usually consisted of handing a stack of punched cards to the operator and hoping for the best. A day later a tense programmer would trot down to the computation center to find out whether the program had compiled or, more likely, that a misplaced comma had thrown the machine off track. Or that a card, instead of having been punched, had only been dimpled.[4] Then the process would have to start over again, before one even got to the debugging stage, let alone to running the program.

But Fejes-Tóth saw far afield. Except for the as-yet undeveloped hardware, the stage was set. Eventually, computers would not only "approximate the minimum of the volume function with great exactitude," as Fejes-Tóth had predicted, but prove Kepler's conjecture with complete rigor. It was to take another quarter of a century, however, until someone picked up the gauntlet. Tom Hales, who was a seven-year-old kid, still playing with Lego blocks when *Regular Figures* appeared, would be the man to put computers to work on Kepler's conjecture.

Thomas Callister Hales was born in 1958 in San Antonio, Texas, and grew up in Provo, Utah, a clean, pleasant city, beautifully situated between the Wasatch Mountains and Utah Lake. His father, Robert Hyrum Hales, was an ophthalmologist (a profession of some importance to the solution of Kepler's conjecture). Hales's grandfather, Wayne Brockbank Hales, was a physicist at Brigham Young University (BYU) in Provo. He was to have a profound influence on Hales's education.

Wayne B. Hales was born in 1893. In the course of his thesis work at the University of Chicago, the University of Utah, and the California Institute of

---

[4] A problem that was still with us in the year 2000, as the presidential showdown between George W. Bush and Al Gore demonstrated.

forty-four or even more sides), Delaunay simplices decompose space into partitions of the same form—four-sided tetrahedra.

By the way, the partitioning of space is not an activity limited to mathematicians. It has applications across many disciplines. A recent book on the subject lists nearly two dozen areas in which Delaunay and Voronoi decompositions of space play a role: anthropology, archaeology, astronomy, biology, cartography, chemistry, computational geometry, crystallography, ecology, forestry, geography, geology, linguistics, marketing, metallurgy, meteorology, operations research, physics, physiology, remote sensing, statistics, and urban and regional planning.

So Hales decided to try his luck with Delaunay simplices. The first step consisted in joining the centers of the spheres with each other, separating space into tetrahedra. But Hales was not satisfied with decomposing space, he also wanted to build something, as he used to do when he played with Lego blocks. The only construction rule he set himself was that all Delaunay tetrahedra had to be attached to a common center. He called the resulting structure a *Delaunay star* (which we shall sometimes abbreviate to D-star).

I mentioned previously that forty-four neighboring spheres could surround a central sphere. By Euler's formula, the Delaunay triangulation of this cluster has eighty-four simplices (see the appendix). Hales showed that Delaunay stars could be made up of at most 102 simplices, which implies

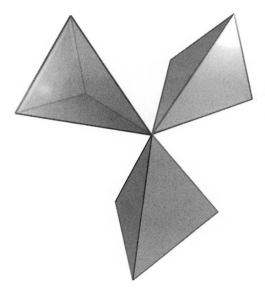

Delaunay star

that up to fifty-three spheres can crowd around a common center (again, see the appendix). Following Fejes-Tóth's hint, Hales was going to study the density of local configurations. But instead of computing the density of V-cells, as had been fashionable for the previous thirty-five years, Hales analyzed the density inside D-stars.

The Delaunay method opened a new avenue for the study of sphere packings, but it was a risky business. The V-cells method wasn't able to prove Kepler's conjecture, but at least it had allowed Claude A. Rogers, John Lindsey, and Doug Muder to provide a series of gradually improving bounds on the packing densities. The hope that some day a bound on the density of 74.05 percent would be reached had remained intact. But D-stars were quite a different matter. There were no intermediate way stations. You either get all the way to the final result or you don't get there at all. There was no glory for the winner of the half-lap.

At the beginning, Hales didn't get there at all. All he managed to show was the following: eight balls placed at the corners of a regular octahedron cover 72.09 percent of an octahedron's volume. That much was standard knowledge. He then applied this knowledge to D-stars and went on to prove that one can not pack more than four additional balls in the remaining 27.91 percent. This was new, but it was not very surprising. In fact, it seemed rather obvious that only so many balls could be packed into such little space. Hales's theorem was a very weak theorem indeed. So one should not be surprised that the consequences of this statement would not turn out to be very satisfying.

It came as it had to. When he applied his new approach, Hales was only able to show that the density of any packing was less than infinity. Infinity? He did not even prove that the density of any packing was less than 100 percent, which would have been quite a useless statement in its own right but, at least, would have made some sense. So Hales's first finding was completely, totally, utterly useless and had absolutely no value in proving a new bound on the density of sphere packing. But his seemingly useless theorem did do one thing: it proved that densities could not be negative. That would have been weirder still. So Hales's early foray into Delaunay triangulations simply set the stage. It demonstrated that the method did not descend into the depths of weirdness—even though it remained, for the time being, within the realms of uselessness.

In January 1990, Hales submitted an article about these findings, entitled "Remarks on the Density of Sphere Packings," to the journal *Combinatorica*. But before it could be accepted, he found what every mathematician dreads: a counterexample to one of his conjectures. Hales had to sit down and revise the paper. Two years later, in December 1991, it was accepted. It then took another two years to make it into print. Even thought it was of little practical use, it proved revolutionary. Hales's "Remarks" suggested a novel approach to Kepler's conjecture. Since the Delaunay approach cannot produce negative densities, there was hope that it could eventually be worked into something useful. For the time being, however, Hales was not able to take the idea any further.

But a seed had been planted. In fact, Hales started to feel a bit cocky. He believed that with some more twisting, squeezing, and tweaking, the Delaunay method could be developed into a strategy to prove Kepler's conjecture. He talked to several colleagues and told them about his ideas. Among the colleagues were Robert Langlands and John Milnor, two very highly regarded mathematicians at the Institute of Advanced Studies (IAS) at Princeton. They listened carefully as Hales laid out his plans, but were not convinced. The two seasoned professors feared that their young colleague would be working his way down a blind alley. They suggested he first test his ideas on some examples, using a computer, to gain some confidence in this novel approach. Only if the tests produced no counterexamples—no D-stars whose density bound rises above 74.05 percent—would it be worthwhile to embark on a time-consuming and possibly not very fruitful expedition. Remember that there were no laurels to be won on the way to the finishing line. Either the Delaunay approach would lead to complete success, in the form of a proof of Kepler's conjecture, or it wouldn't even win Hales a booby prize in the form of, say, an improved upper bound on the packing density. Everything would have been in vain. (At the same time, Wu-Yi Hsiang was also visiting the IAS. Langlands invited the two men to his office, where they had their first friendly talk about Kepler's conjecture.)

Hales took his colleagues up on their suggestion. He wrote a computer program and began to test his hypothesis. After running the program on many, many clusters he found that none of them had a density bound above 74.08 percent. Wait a minute—74.08? To Hales's chagrin, the program had identified a cluster with a density bound just slightly greater than Kepler's 74.05 percent. Imagine his surprise. Could it be that a packing arrangement existed after all that was denser than Kepler's packing? An arrangement that had not been discovered until now? This seemed extremely unlikely, so Hales's first thought was that his computer program contained a bug. He printed out a listing of the program and painstakingly double-checked the code. He went over it line by line, examining every comma and every semicolon, but found no mistake. There was nothing wrong with the code.

Back to the results. Hales subjected the offensive sphere arrangement to more scrutiny. After a while he had to admit that the cluster he had found was, in fact, a genuine counterexample to his hypothesis. It consisted of a central sphere, surrounded by a dozen neighbors. But the neighbors were arranged differently than in Kepler's packing. One ball was on top of the central sphere, five balls were wrapped around it, just above the equator, another five were wrapped around it, a bit below the equator, and finally one ball was located at the bottom.[6] When a V-cell is formed around the central ball, it has the shape of a polyhedron with one pentagon on top, another one at the bottom, and ten triangles around the middle. The arrangement is called the pentagonal prism. We shall call it the *dirty dozen*. It would haunt Hales for many years to come.[7]

The existence of this bothersome cluster did not mean that Hales had actually found a better packing. It just meant that the true density lay somewhat below 74.08 percent. His suggested method would have to be improved. For the time being, Hales decided to collect everything he had

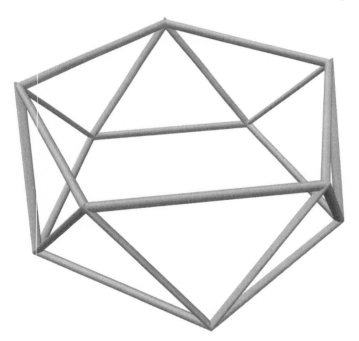

Pentagonal prism (dirty dozen)

---

[6] We discussed this so-called icosahedral arrangement in chapter 5.
[7] Since the twelve surrounding spheres can be moved slightly, there are actually infinitely many *dirty dozens*.

found out so far. He wrote a paper entitled "The Sphere Packing Problem" and submitted it to the *Journal of Computational and Applied Mathematics*. It was published in 1992, shortly after he had finished revising his previous article.

Kepler's conjecture never loosened its grip on Hales. At the beginning of 1994 he decided to put everything else aside and spend all his available time on the problem. He was by then near the top of his profession. He had just ended a three-year appointment at the University of Chicago and had been invited as a visiting member to the IAS, one of the most, if not *the* most prestigious academic institution in the United States. Hales did not have to worry about academic appointments. To a young mathematician of his stature, tenure would certainly come sooner or later. He could afford to invest a year's worth of his time and energy tinkering with a question that may possibly turn out not to be solvable with the methods he was proposing.

Hales spent most of the year dismantling and reassembling Delaunay stars. He had by now been playing around with the tetrahedra for a longer time than he had spent with Lego blocks as a kid: more than five years. But all his struggles came to naught. Toward the end of the period he had reserved for Kepler's conjecture, he became a bit discouraged. He had found a procedure to partition space into fragments and to compute their densities. But whenever one of the pieces was too large, his method gave inaccurate results. It became extremely difficult to prove anything about the sphere packing. The problems seemed insurmountable.

Suddenly, in November of that year, inspiration came, seemingly out of nowhere. Hales was sitting in on one of the weekly seminars at IAS. Robert MacPherson, a permanent member of the institute, was holding forth on the question of why the French mathematician Jean Leray (1906–1998) had not discovered *perverse sheaves*. It is a fascinating subject to the initiated, and Robert MacPherson was just the person to give a lecture about it. Two years earlier he had received an award from the National Academy of Sciences for his pioneering role in the introduction and application of radically new approaches, among them perverse sheaves, to the topology of singular spaces. But for everyone else sheaves, perverse or otherwise, are a major bore. While MacPherson was giving his overview, Hales started to daydream about Voronoi and Delaunay decompositions. All of a sudden, MacPherson's voice cut through the fog of his reverie. "But, of course, there are more than two ways to do things." It hit Hales like a lightning bolt. That was it! Voronoi cells and Delaunay simplices cannot be the only techniques to partition space. There must be more than two ways to do things. Hales later recalled: "Bob's statement jolted my daydreams so forcefully that I began a serious investigation of other decompositions of space."

The idea that would eventually lead to a solution of Kepler's conjecture came to Hales in a dream the very night after the seminar: If neither method works, why not try using them both? Maybe the advantages of partitioning space into Voronoi cells and Delaunay simplices could be combined into a hybrid approach?

Let's take it step by step. One characteristic of a good packing is a small Voronoi cell. But this does not suffice. Half a century earlier, Fejes-Tóth had stated that even though the regular dodecahedron minimizes the volume, the cells of the next outer layer waste too much space. To circumvent this problem, he suggested that the cell and its neighbors be taken into account simultaneously. He proposed assigning a score to each cluster of balls. One requirement for a useful score should be that the volume of the central ball's V-cell be small. The other requirement should be that the ball belongs to a tight cluster. To combine both requirements, the score must be designed in such a way that it decreases if the ball is part of a "bad" arrangement. Hence, an inefficient arrangement would receive a bad (low) score, even if the V-cell were small. On the other hand, a tight packing would be awarded a good (high) score, even if the central ball's V-cell were a bit larger.

Hales's first task, therefore, was to devise a workable penalization schedule. He decided to design a function of the form *Volume + Penalty* that would be minimized. The penalty could be positive, zero, or negative. Once a suitable score had been decided upon, Hales would check to see for which arrangement *Volume + Penalty* achieves its minimum. The expectation was, of course, that this would happen for Kepler's sphere arrangement.

Sometimes mathematicians choose to *maximize* the reverse function instead of minimizing the original function. It is like screaming at a taxi driver, "drive as fast as you can" instead of yelling, "spend as little time as you can to get there." Both of these polite requests amount to the same thing—especially if you back them up with a $10 bill. Hales did something similar. He *maximized* the *negative* of the *Volume + Penalty* function.

In his search for an appropriate function, Hales first established a baseline. The regular octahedron, with edges of length 2 and with one ball placed at each of the six corners, would serve that purpose. The parts of the eight balls that are contained within the octahedron fill 72.09 percent of the volume. Henceforth, all arrangements would be measured against that standard. A cluster of balls that is less dense than the octahedral arrangement would be slapped with a penalty, and a cluster that is denser than the octahedral arrangement would receive a bonus. The octahedron itself would receive a score of zero. In the original versions of his papers Hales called this construct the *Gamma-*

function. Since this was not a very illustrative name, friends suggested that he give it a designation more indicative of its purpose. So Hales called it the *compression,* which does not really convey much more information. We shall call it the *surplus-function,* because it represents the surplus of the specific arrangement over and above the octahedral arrangement.

Let us examine an example. Balls placed at the corners of the regular tetrahedron fill 77.96 percent of the tetrahedron's space. The difference to the octahedron's density is 5.87 percent (= 77.96% − 72.09%). Since the volume of a tetrahedron with edge-length 2 is 0.942, the surplus awarded to the tetrahedron is 0.0553736.[8] Hales called this number a *point (pt).* So 1 pt defines the "density surplus" of a tetrahedron over an octahedron. The surplus of all cells, stars, and simplices would be measured in points (pts). If the balls in the simplex fill *less* space than in an octahedron, the surplus would be negative.

But soon a problem arose. Whenever the simplex was large the computations of the surplus were fraught with errors and gave useless results. Hales was at a loss. He thought about simplices, then about cells, and then about simplices again. And then it hit him: the thing to do was to use a combination of both simplices and cells. Whenever the simplices were small, he would compute the surplus of the simplex, as he had done until now. But when the simplex is larger than a certain cutoff value, he would switch to the V-cell of the central ball and compute its *premium.*

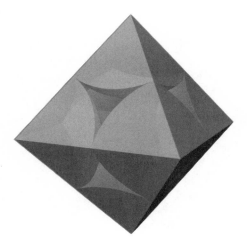

Octahedron used for surplus–function

---

[8] The volume of a tetrahedron with edge length k is $\sqrt[1]{2}\,k^3$. For $k = 2$ the volume amounts to 0.9428. The difference in densities between the tetrahedron and the octahedron, 5.87%, times the volume of the tetrahedron, 0.9428 equals 0.0553736.

What's a premium? The premium is to cells what the surplus is to simplices. Take the central sphere and consider its V-cell. Slice the cell into wedges. The bottom corner of each wedge is occupied by the sphere. Now compare the density of such a wedge to the density of the octahedron. The premium is the wedge's gain in volume over the octahedron. The *score* of a Delaunay star would then be the sum of all the surpluses and premiums. That was Hales's brainstorm.

The crucial question now was: What is the maximum score of a Delaunay star? Hales had a very precise idea of what the highest score could be. He was convinced that no star could ever score more than 8 pts. And this, ladies and gentlemen, implies Kepler's conjecture. Why? Because Kepler's arrangement is made up of stars that consist of eight tetrahedra and six octahedra. By definition, octahedra, which have no surpluses, score zero pts. A tetrahedron, on the other hand, has a surplus of exactly 1 pt. Hence Kepler's cluster scores 8 pts. If it could be shown that all other stars score less than that, Kepler's conjecture would be proven.

So all that remained was to show that no matter how space is divided into cells and simplices, no star could ever reach a score higher than 8 pts. Sounds straightforward enough, but don't underestimate the problems. Hales's task resembled an extremely intricate and convoluted three-dimensional puzzle whose pieces—on top of everything else—could penetrate, pierce, and perforate each other.

❖ ❖ ❖

Hales restricted his attention to clusters made up of a central ball and all its neighbors.[9] He drew lines between the centers of the neighbors and projected these lines onto the shell of the central ball. This defined a network

Premium of a V-cell

<hr />

[9] All balls whose centers lie within a distance of 2.51 from the central ball's center were defined as neighbors. Of course, Hales also had to take into account the next layers of spheres (the neighbors' neighbors) in order to define the outside walls of the first layer's V-cells.

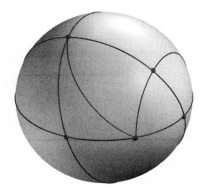

Network on central ball

on the shell of the central ball, which is reminiscent of Reinhold Hoppe's and John Leech's nets from chapter 6. During the following years, Hales would keep himself busy considering all possible such nets. First he would generate them using combinatorial techniques. Then he would classify them according to whether they were made up of triangles, quadrangles, or other polygons. Finally he would compute the maximum score the nets could achieve and show that this maximum lies below 8 pts, except for the FCC and the HCP, which score exactly 8 pts.

The author Simon Singh described Hales's approach succinctly as follows: Plot the density of all 50-ball arrangements on a 150-dimensional graph. Such a graph resembles a 150-dimensional landscape. Construct a 150-dimensional roof over the landscape. Seek the roof's peak. Then lower the roof until it just touches the tallest hill. The "height" of the lowest roof's peak is just 8 pts—if all goes as planned.

This is the general outline of the idea that had come to Hales during MacPherson's lecture and in his sleep in November 1994. He immediately went to work, and started by mapping out a master plan. Hales's strategy was to divide the proof into five segments, which he thought would require roughly similar amounts of effort and time.

The Master Plan[10]

Segment 1: Show that all nets made up of only triangles score at most 8 pts.

---

[10] I have barely scratched the surface of the proof here and will give a somewhat more detailed, step-by-step account of Hales's master plan in the appendix, entitled "The Proof—An Explanation."

Segment 2: Show that three-sided loops score at most 1 *pt,* rectangles score at most zero points, and any loop with more than four sides gets a *negative* score.

Segment 3: Show that all nets made up of triangles or quadrangles (with the exception of the *dirty dozen*) score less than 8 pts.

Segment 4: Show that all nets that contain at least one region with more than four sides score less than 8 pts.

Segment 5: Prove that the *dirty dozen* also has a score of less than 8 pts.

So accustomed was Hales to partitioning space into cells and stars that dividing time into segments must have seemed very natural to him. (Hales' degree in engineering-economic systems also came in handy.) But it was a rather unusual procedure and resembled more an engineer's approach to the building of bridges than a mathematician's approach to proving a theorem. A scientist's progress usually comes in completely unpredictable spurts and starts. But this nonconformist approach would enable Hales to chart his progress and determine whether he was on or behind schedule. At the end of 1994 Hales had already completed a significant part of Segment 1, and at a pace of one segment per year, he figured he would finish his task by the end of 1998.

At this point computers enter the scene, as Fejes-Tóth had predicted. They were used throughout the proof to partition space, generate all possible nets, compute the density of all the bits and pieces, and perform innumerable other tasks.

After several months of work Hales managed to prove that no net could score more than 8 pts if it is composed only of triangles. The first segment of his master plan was completed, and he sat down to write the paper. He entitled it "Sphere Packings I" and, on May 12, 1994, submitted it to the journal *Discrete and Computational Geometry.* The referees were not delighted. Hales had not yet formulated his ideas in a sufficiently clear manner. There was no conjecture that a particular scoring system would lead to a proof of Kepler's conjecture, just a suggestion that a solution may eventually be found. There was no mention of the hybrid scoring system. Instead, Hales proposed a "repacking" scheme that he already knew would present huge obstacles.

The editor asked for a revision and Hales spent the following year rewriting his paper. He submitted the revised version on April 24, 1995. Again, the editors were not enthusiastic. They demanded more changes. Dutifully, Hales spent another year—he was already busy working on Segments 2 and 3 at that time—re-revising his paper. He submitted it on April 11, 1996, and this time it was finally accepted. One year later it appeared in print.

With that the ice was broken; Hales had exposed his plan to the world. Was he worried that someone would steal his ideas and make it to the finishing line before he did? He was not. In fact, he was hoping to get some sort of informal approval for his ideas. The size and difficulty of the remaining problems seemed overwhelming and he looked forward to all the help he could get. (The pentagonal prism—the *dirty dozen*—had not yet been identified as a potential problem.) But there were no takers. His friends and colleagues were generally pessimistic about his chances of success and took a wait-and-see attitude. One of Hales's later papers, in which he outlined his master plan, was rejected as being too tentative and speculative. In fact, John Conway had predicted that the problem would not be solved during his lifetime.

But Hales toiled on. In Segment 2 he related scores to the shapes of the net's loops. So which stars could score exactly 8 pts? We already know that Kepler's sphere arrangements reach the perfect score. Hales showed that any deformation of this arrangement, even just the slightest wiggle, reduces the score. On April 24, 1995, Hales put "Sphere Packings II" into an envelope together with the revision of Segment I and sent the two papers to *Discrete and Computational Geometry*. By that time the editors were already somewhat familiar with his work and demanded only one revision. It took Hales a year—he was simultaneously working on the second revision of "Sphere Packings I"—but in April 1996 both papers were sent off to the editor and duly accepted. "Sphere Packings II" appeared in the following year, 562 pages behind "Sphere Packings I."

On to Segment 3, where all nets are dealt with that are woven into both triangles and quadrangles. Of course, the FCC and the HCP have such a net and score exactly 8 pts. Hales wanted to show that all others must score less than that. The problem of the dirty dozen remained. But at least he was able to show that apart from this single exception, all such nets score less than 8 *pts*. When he sat down to write up his findings he toyed with catchy titles such as "Sphere Packings: The Sequel" or "The Return of the Sphere Packing." In the end he decided against it. He could not afford to be anything less than serious, especially with a subject that not everybody took quite seriously anyway. So Hales gave it the unassuming, but not unexpected, title, "Sphere Packings III."

In "Sphere Packings IV" he showed that even if polygons with more than four sides are woven into the nets, they still score less than 8 pts. "Sphere Packings III" and "Sphere Packings IV" were never published. Hales was just happy to know that he was on the way to solving the problem.

Now all that was missing was Segment 5. The pentagonal prism had been a major thorn in Hales's side ever since he had come across it during his computer experiments at IAS. But now, at long last, Hales was going to receive some help. At this point his father's profession acquires importance for our narrative. Dr. Hales had a well-established eye clinic in Provo, Utah. One of his patients, Helaman Pratt Ferguson, was a professor of mathematics from Brigham Young University.

Ferguson is one of the few lucky individuals who are able to combine two passions and excel at both of them. His early childhood was sad. A lightning bolt killed his mother while she was hanging laundry in the back yard, and then his father was drafted into the army in World War II. The little boy and his sister were taken in by their grandmother. But when horrified relatives found out that the old lady had been serving them coffee, they arranged for their adoption by distant relatives. Ferguson started his training as an apprentice with his adoptive father, an Irish stone mason, and then went on to study painting and sculpture. But since not every aspiring artist can be certain to make a living, Ferguson also studied mathematics. In 1971, he received his Ph.D. from the University of Washington in Seattle and then taught mathematics for seventeen years at BYU. He also did research on computational number theory and one of his computer algorithms became famous. It was the so-called *PSLQ*-algorithm, which permits the calculation of the nth digit of the number π without calculating the $n - 1$ digits before it. *PSLQ* also does some other neat things for mathematics and physics, and a professional journal named it one of the top ten algorithms of the century.[11]

One would think that such an achievement would have sufficed for a man's lifetime. Not so for Ferguson. He is, above all, a sculptor. His mathematically inspired creations, which combine esthetic beauty with mathematical elegance, are exhibited all over the world. Ferguson manages to translate rigorous concepts into three-dimensional art, and thus give physical expression to the beauty of mathematics.

Ferguson has yet another claim to fame. To keep in shape, he took up jogging. But after pursuing this activity for more than two decades, he started to find it a bit lopsided. The arms flail about uselessly in the air while the legs do all the work. Ferguson wanted to give his arms something to do, so he took to "joggling." He juggles rings, balls, or bowling pins while he jogs. Apart from exercising his arms, says Ferguson, joggling also takes the boredom out of jogging. The *Guinness Book of Records* lists Helaman P. Ferguson as the first person to have joggled for a length of 80 kilometers. (It is

---

[11] To be exact, the algorithm discovered the formula that produces a hexadecimal expansion of π. Another mathematician, David Bailey, also from BYU, wrote the code for *PSLQ*.

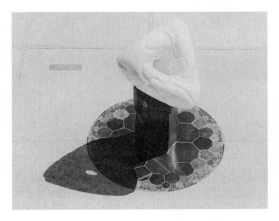

A Helaman P. Ferguson sculpture

not recorded whether he covered the distance without dropping a ball.)
Unfortunately, he had to stop joggling after he injured his back in a fall. To
this day, however, he continues to "wuggle" (walk and juggle).

Despite his many skills, Helaman Ferguson was not the one who would
aid Hales in his struggles. This task was reserved for one of his sons, Samuel
Lehi Pratt Ferguson. Sam Ferguson was Hales's junior by eight years. They
had known each other in Provo, and Ferguson had been invited to IAS in
1983, when Hales was doing his doctorate at Princeton. The young graduate student came over occasionally to visit the Ferguson family and Sam
got to know him a little better.

Early on in life Ferguson showed interest in science. After his father
and mother had convinced him that mathematics was the key to the other
sciences, he took up the subject seriously. It seemed to be the best use of
his talents. (Ferguson does not describe himself as an outstanding mathematician. He thinks of himself as more of a problem solver, but a pretty
good one.) He attended BYU, where he received his B.S. degree in 1991.
In his senior year, Hales came for a visit to the mathematics department
and gave a lecture in which he discussed his work. Ferguson was in the
audience. This was the first time he heard about Kepler's conjecture.

After graduation, Ferguson took his time to decide whether to continue
with his studies. When he finally decided to apply for graduate school, the
deadlines of many universities had already expired. One graduate department
that had not yet closed its doors was at the University of Michigan. So Ferguson sent in his application and was accepted. His second year there coincided with the arrival of his old friend, Hales. The University of Chicago had
been too coy to offer Hales tenure, but the University of Michigan did not

Samuel L. P. Ferguson

require much prompting. An offer was made and Hales accepted. At Michigan, Ferguson and Hales got along very well, which wasn't surprising given their common youth in Provo and their shared Mormon background.

Ferguson got his Master's degree in 1993 and decided to stay on to do his doctorate. Unfortunately—with hindsight, we may say fortunately—his search for a thesis advisor coincided with the remodeling of Angell Hall, where the department of mathematics was housed. The ensuing chaos encouraged all faculty members who could leave to do so. Ferguson had a hard time finding a thesis advisor. One of the professors who was not yet up for a sabbatical leave and had to stay put amidst the dust, dirt, and grime, was none other than Tom Hales. After consulting with his father, Ferguson decided to ask Hales to be his thesis adviser. Nothing seemed more natural than that the two Provonians would collaborate in order to push Kepler's conjecture to the finishing line. At a time when most seasoned mathematicians were still very skeptical about Hales's master plan, Ferguson was just naïve enough to believe that it could work. Hales recounted his problem with the dirty dozen and asked him to "make it go away." And that's exactly what Ferguson did. The dirty dozen became the subject of his doctoral thesis. His talents as a problem solver served him well during the long and arduous path that lay ahead. To help his son with visualization, Ferguson senior sent him a large number of ball bearings. Made of solid steel, about 1 inch in diameter, they were a very impressive prop. Unfortunately Ferguson could not really use them as his father had intended, since they were too heavy to be kept in place with chewing gum.

Originally, Hales thought it would take Ferguson a few months to complete Segment 5. He would then give his student something else to cut his teeth with, in order to fatten up the thesis. But the amount of work turned

WOBBLY BALLS AND HYBRID STARS   179

out to be significantly more than expected and took much longer than just a summer. In fact, Ferguson ended up spending three years to complete it. The programming and the computer experiments he and Hales designed to find the proper decomposition and scoring system were a tremendous amount of work. In the end, there was certainly no need to add material to beef up the thesis.

One day, in August 1997, the work was over. Segment 5 was solved. Kepler's conjecture had been proved. Were they elated? Not really. Hales describes the sensation he felt at that moment more as a sense of relief. It was as if a weight had been taken off his shoulders.

Ferguson reported the findings in "Sphere Packings V." The thesis was accepted and Sam, now Dr. Ferguson, wanted to submit it for publication to a respectable journal. Hales thought that the journal *Discrete and Computational Geometry* would be the right outlet, since this is where he had published his first two papers in the series. He talked to the editors but they were not interested. It wasn't a solution to Kepler's conjecture, they said, merely partial progress towards that end. As such, it could not stand on its own. On top of that, "it would be quite difficult to find a referee willing and able to dig through the sordid details of the proof," as Ferguson himself admits. To this day his paper is still waiting to appear in print. Ferguson is now working as a mathematician with the U.S. Department of Defense.

Ferguson describes his collaboration with Hales as an extraordinary experience. Both benefited from the partnership. It helped Hales a great deal to have someone working on the problem together with him. It would have been much easier to put the project on the back burner, save for the fact that Ferguson's thesis was hanging in the balance.

In the winter of 1998 Hales was nominated for the coming year's Henry Russell award of the University of Michigan. The award is conferred annually to a junior faculty member in recognition of distinguished scholarship and conspicuous ability as a teacher. Ferguson wrote an eloquent letter in support of Hales's nomination and Hales won the Henry Russell award in 1999.

To sum up, nets with triangles were treated in "Sphere Packings I." Nets with triangles and quadrangles were treated first in "Sphere Packings III" (Kepler's arrangement and all deformations), and second in "Sphere Packings V" (the dirty dozen). Nets with polygons that have more than five sides were dealt with in "Sphere Packings IV." Thus all possible nets were covered. ("Sphere Packings II" was needed only as "supporting material.") Every single net scores less than 8 pts, except for the net of Kepler's arrangement, which scores *exactly* 8 pts.

Helaman Ferguson once asked Hales why he chose to tackle Kepler's conjecture. Hales answered that he initially thought it would not be that hard to solve. Had he known how hard it would be, he said, he would not have embarked on it. This expresses in a nutshell the obstacles and hurdles that had to be surmounted. The next two chapters describe in a little more detail the work that went into the five segments.

Before we end the chapter, I need to set the record straight on one point. One might come away from this chapter with the impression that Fejes-Tóth was a bumbling dreamer whose work mostly contained unfulfilled promises and unproven hypotheses. This does not represent the whole picture; We must see his contributions in perspective. As Hales put it:

> I admire Laszlo Fejes-Tóth a great deal. During my long years on the problem, I felt that he, more than anyone else, understood the nature of the problem and had covered the terrain before me. On many occasions, after arriving at various insights after long and hard reflection, I found evidence in his work that he had made the journey already and had arrived at the same insight before me. That he had to formulate so much as hypothesis and conjecture only shows that he was a visionary working on the problem 50 years before its time. I admire him for that in the same way that I admire any great mathematician who formulates a bold proposal to solve an important problem.

CHAPTER 12

Simplex, Cplex, and Symbolic Mathematics

The proof of Kepler's conjecture is basically an optimization problem. Tom Hales had reduced the proof to the task of optimizing the score of Delaunay stars. Kepler's configurations score 8 pts, and Tom showed that no star could score more than that. His task had consisted, firstly, of finding the maximal score of all possible stars and, secondly, of showing that this maximum score lies below 8 pts. One hurdle was the listing of all possible Delaunay stars.[1] Another was the maximization of the scores.

Maximization problems arise all the time, as we constantly strive to maximize something: income, grade-point average, profit, strength, speed, pleasure. On other occasions one may want to minimize variables, like expenditures, effort, the distance one has to walk, pain, or load. Maximization and minimization problems are subsumed under the notion of "optimization." Maximizing profits may, for example, be equivalent to minimizing costs.

The answers to many optimization problems may reduce to zero or infinity. For example, how many doodads should a factory produce in order to minimize the cost? Well, zero. Do nothing and you won't have to pay anything. But this answer is, of course, too simpleminded. One wants to minimize costs, but one also wants to produce *something*. So, in general, an optimization problem takes the form of "maximize profits, given that at least twenty-five doodads are produced." The "given that" clause is called a constraint. Only constrained optimization problems are of interest. Unconstrained problems have so-called trivial corner solutions (zero and infinity are considered to lie in the corners of infinite space).

---

[1] Or, equivalently, their nets.

The earliest progress on constrained optimization was made in the 1930s by the mathematician Leonid Vitalievich Kantorovich in the Soviet Union. This is not altogether surprising; after all, the proponents of central planning were convinced they could maximize everybody's welfare by having supercomputers (which, alas, were not yet available) decide on optimal production quotas, allocation of materials, pricing schemes, and consumption schedules. As befits a centralized economy, Big Brother would take care of everything. As befits the Western world, Russian research was completely ignored.

Kantorovich was born in 1912, in St. Petersburg (later Petrograd, then Leningrad, and then St. Petersburg again). His earliest childhood memories included bullets whizzing around during the October Revolution of 1917. At age fourteen he entered the mathematics department of Leningrad State University. He wasted little time and got his Ph.D. when he was only eighteen years old. Kantorovich was a theoretical mathematician with an excellent feel for the mathematics underlying economics. His involvement with economics started by accident. As a twenty-six-year-old professor whose salary was not exceedingly high, he moonlighted as a consultant to a factory that dealt with plywood. He was asked to determine the distribution of raw materials that would maximize the productivity of the equipment. Kantorovich solved the problem by inventing a procedure that he called the "method of resolving multipliers." In 1939 Leningrad University Press published a booklet by Kantorovich that contained the main ideas of the theories and algorithms of what would later become known as linear programming. Written in Russian, his work remained unknown to Western scholars for decades.

At about the same time a Hungarian émigré, John von Neumann, was doing research at the Institute for Advanced Studies (IAS) at Princeton. In fact, it is said that IAS was founded in order to give him and his colleague Albert Einstein a place to further human knowledge. Von Neumann was born 1903, the son of a wealthy Jewish banker in Budapest. Jancsi, as he was called then, was a child prodigy and despite a strict quota against Jewish students at the University of Budapest he was easily admitted. But he paid back in kind, and hardly ever attended any classes. Instead, he simultaneously enrolled at the University of Berlin. His attendance record at Berlin was not outstanding either, because in 1926 he received a diploma in chemical engineering from the Federal Institute of Technology (ETH) in Zürich. In the same year he obtained a doctorate from the University of Budapest. There followed a study year with David Hilbert in Göttingen. Everyone who met him recognized his superior intellect. During the years 1930 to 1933 he held positions both in Germany and at Princeton University. After

IAS was founded, he became one of its six original professors of mathematics. In 1937 von Neumann, now Johnnie, became an American citizen. He died of cancer in 1957, at age fifty-four.

Von Neumann is considered the father of modern computers. In fact, he fathered many disciplines and subdisciplines. His contributions to mathematics, quantum theory, economics, decision theory, computer science, neurology, and other fields are too vast to be listed here. We'll just mention two areas. As a consultant to the Manhattan Project, he was instrumental in the development of the atom bomb in Los Alamos. He worked out the theory of "implosion," which proved to be the key to the developments of the bombs Little Boy, dropped over Hiroshima, and Fat Man, dropped over Nagasaki.[2]

At the same time von Neumann was also preoccupied with something much less sinister. Together with his Princeton colleague Oscar Morgenstern, an émigré from Austria, he was working on a fun paper about games people play.[3] When it turned out that the material would be too much for a single paper, the two professors planned to cover the subject in a series of papers. Then it turned out that even a series of papers would not be able to contain everything they had to say about games. So they decided to write a book. The fruit of their labor appeared in 1944 as a thick primer that would become one of the most influential scientific works of the twentieth century. *Theory of Games and Economic Behavior* ushered in the age of mathematical economics.

The best-known game of the type von Neumann and Morgenstern studied is the so-called prisoner's dilemma. Two suspected burglars have been arrested and are being interrogated separately. The police do not have sufficient evidence against either of them, so they try to make a deal. If one of the suspects is willing to turn state's witness, he will get off scot-free. The other will be convicted based on the testimony and receive a sentence of two years. If both of them keep their mouths shut, they will be convicted on a minor charge and each get a sentence of one week. If, however, unbeknownst to each other, both of them spill the beans, they will both receive a sentence of six months. Now put yourself into the shoes of a prisoner. If you and your buddy keep your mouths shut, you'll both be out in a week. If you keep your mouth shut, but your buddy—who isn't such a buddy,

---

[2] It is for his association with the Manhattan Project (and for the fact that the cancer-stricken professor was confined to a wheelchair during the last months of his life) that Stanley Kubrick reportedly had von Neumann in mind when he created the character Dr. Strangelove in his 1963 film.

[3] The origins of game theory go back to the nineteenth century, to Augustin Cournot and Francis Edgeworth, and to the 1920s to Emil Borel.

after all—starts singing, you'll be in the slammer for two full years. If you talk and your buddy—who doesn't yet know that you're no buddy of his— keeps quiet, you're out, and he'll be making license plates for two years. If both of you talk to the police, you'll both be doing time, although only for half a year. The prison sentences are set out in a so-called payoff matrix. What should you do?

	You Talk	**You Don't Talk**
Buddy Talks	Both of you get six months	You get two years and buddy goes free
Buddy Doesn't Talk	Buddy gets two years and you go free	Both of you get one week

There is no "correct" answer. The dilemma is precisely that: no prisoner can make the "correct" decision without knowing what the other will do. The so-called rational strategy is to make a deal with the police. But what if you two hoodlums plan further burglaries? Will you be able to count on each other in the future? This game is called the *iterated* prisoner's dilemma. The best approach for serial burglars is a tit-for-tat strategy: always cooperate (with your mate, that is, not with the police), except if your mate doesn't. If he defects, punish him in the next round by turning state's witness. That'll teach him. The tit-for-tat strategy has been suggested as a possible explanation how cooperation evolved in a world where, according to accepted wisdom, everybody is supposed to be selfish. In 1994 the Americans John C. Harsanyi and John F. Nash received the Nobel Prize for economics together with Reinhart Selten from Germany for their work in the theory of games.[4]

The first major breakthrough in game theory was the "minimax theorem" that von Neumann discovered in 1928, about ten years before he met Morgenstern. Let us say two players compete for the same amount of money. The winner's gain is the loser's loss. This is called a zero-sum game, because the sum total of all winnings and losses is zero. The minimax theorem states that

---

[4] The fascinating biography of John Nash, who spent a quarter of a century as a paranoid schizophrenic, involuntarily confined for long stretches of time to mental institutions before miraculously recovering, is related in the play *Proof* by David Auburn, in the book *A Beautiful Mind* by Sylvia Nasar, and in the Oscar-winning movie of the same title directed by Ron Howard.

the problem of maximizing the minimum gain has the same solution as the problem of minimizing the maximum loss. To explain, say it's your lucky day and you are convinced you will win. But you will win different amounts of money depending on which strategy you chose. A good approach to the game would be to check the worst scenario in each strategy. Then choose the strategy that guarantees you the largest payoff, even if the worst happens. In other words, *maxi*mize the *mini*mum gain. On an unlucky day you may want to employ a different tactic. You feel like you will lose no matter what you do, but you'd like to keep the damage to your wallet to a minimum. So check the worst-case scenario in each strategy, and then choose the strategy that *mini*mizes the *maxi*mum loss. Surprise, surprise—in a zero-sum game these two strategies coincide. Maximin equals minimax.

A few years after von Neumann discovered the minimax theorem, it became apparent that it had an exact counterpart in Kantorovich's theory of linear programming. This counterpart is called "duality theory." Let's say a doll factory produces soldier dolls and dress-up dolls. Plastic, textiles, and color are required for both dolls in differing amounts. Given the prices of the dolls and the amounts of raw material that are available, the factory wants to maximize profits. What quantities of soldier dolls and of dress-up dolls should the factory produce? That is the primary problem.

Now on to the dual. The dual problem consists of interchanging the profit function and the constraints and then solving that new problem.[5] The answer to the new problem indicates how profits would increase if the factory had another unit of raw material at its disposal. It gives the maximum prices that the factory would be willing to pay for additional amounts of plastic, textiles, and color. It may be a bit difficult to grasp why the dual is equivalent to the primary problem, but that's why duality theory is so surprising.

With the proof that the minimax theorem is equivalent to duality theory, von Neumann had provided the theoretical underpinnings for linear programming, which is the method Hales would be using in his proof of Kepler's conjecture. Kantorovich, for his part, had suggested a practical method for the solution of optimization problems. It did not utilize differentiation or anything as complicated in order to find optimal solutions. All that was required was a hop, step, and jump to higher-dimensional space. But Kantorovich was ahead of his time. Computing machines—which would have been able to deal with large-scale hops, steps, and jumps—did not exist yet. That started to change in the late 1940s when the first computers were built under von Neumann's direction. They were designed to

---

[5] This is done by inverting the rows and columns of a linear programming matrix.

handle the massive calculations and repetitive tasks that were necessary for optimization problems. All that was missing was a suitable algorithm.

At this point the action moves back to the other side of the world. A few years after Leningrad University Press published Kantorovich's book, and unbeknownst to the Russian professor, efforts were underway in the U.S. Air Force to allocate resources efficiently to different tasks. It was 1947 and the war had recently ended. George B. Dantzig, a thirty-three-year-old statistician, was working in the Pentagon on training and supply schedules.

Dantzig was born in Portland, Oregon, in 1914, completed his undergraduate studies at the University of Maryland, and then received an M.A. degree from the University of Michigan. Upon graduation, he worked as statistician in the U.S. Bureau of Labor Statistics. He then joined the war effort and from 1941 to 1946 was head of the Combat Analysis Branch, U.S. Air Force Headquarters Statistical Control. In the Air Force he made a name for himself as an expert on programming—planning methods done with desk calculators. (Program, at the time, was a military term that referred to plans and schedules for training, supply, and deployment of men.) Dantzig earned a doctorate in 1946 from the University of California at Berkeley and was appointed mathematical adviser to U.S. Air Force Headquarters in the same year. His assignment in this new position was to devise more efficient planning methods.

He rose to the challenge and invented an algorithm called the *simplex*. It was listed by a professional journal as one of the ten most important algorithms of the twentieth century, and is still used today in many different variants.[6] In fact, one expert estimated that the simplex algorithm accounts for more computer time all over the world than any other program, with the possible exception of database handling.

The inner workings of this algorithm can be easily demonstrated on a sheet of paper. Let us look at the doll factory again. It has two products, soldier dolls and dress-up dolls; there are constraints on the raw materials; and the objective is to maximize profits. We plot the number of dolls along the axes, soldier dolls in one direction and dress-up dolls in the other. Then we add the constraints; these are just straight lines. All combination of soldier and dress-up dolls on one side of the line are possible. The points on the other side represent combinations that are infeasible. They may be using too much plastic, for example, thus violating a constraint. The area below and to the left of all the constraints defines the *feasible region*, that is, the combi-

---

[6] We met another one of the top ten in chapter 11: Helaman Ferguson's PSLQ algorithm.

nation of values that fulfill all restrictions. We seek the optimal point in the feasible region, which is tantamount to pushing a suitable profit line as far as possible. Dantzig realized that the optimal combination of soldier dolls and dress–up dolls must lie on an edge of the feasible region. If one chooses an interior point, it will always be possible to find another combination of dolls, situated closer to an edge, with increased profits. For similar reasons the optimal solution can only lie at a corner, that is, at a point where two constraints meet. With this, the stage for the simplex algorithm is set. The computer starts with one corner, moves as far as possible along an edge in the direction in which profits increase fastest. Having arrived at another corner it pivots around again and starts moving along the new edge. At the next corner it pivots again, and thus the process continues. When no further increase in profits is possible in any direction, the algorithm stops: it has reached the optimal combination of soldier dolls and dress–up dolls.

The simplex algorithm can be applied even if the factory has a broader product palette and operates under many more constraints. Instead of moving from corner to corner on a sheet of paper, the simplex algorithm roams around from corner to corner in higher-dimensional space. Problems with thousands of variables and constraints are now routinely solved every day.

Dantzig left the Pentagon in 1952 and became a research mathematician with the RAND Corporation. In 1960 he was named professor of computer science at Berkeley and six years later he moved to Stanford University. He never won a Nobel Prize in economics even though many colleagues believed that he richly deserved one. Instead, Kantorovich and Tjalling C. Koopmans, a Dutch–born U.S. economist, shared the honor in 1975, for their work in the theory of optimum allocation of resources. This

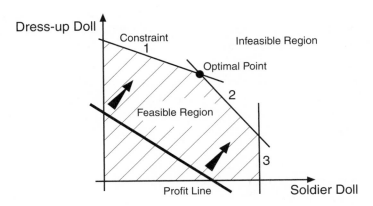

Linear program in two dimensions

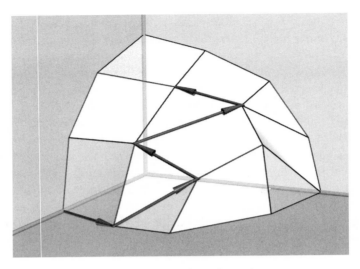

Linear program in three dimensions

was, in effect, an application of linear programming to economics. Koopmans, who was a good friend of Dantzig's, at first considered refusing the prize. He was convinced by a fellow Nobel laureate to accept in spite of his misgivings, but could not quell his conscience. After he had received his half of the $240,000 prize money, he donated $40,000 to a research institute, thus reducing his share to $80,000, which is the amount he would have received had Dantzig been the third cowinner. Even though Koopmans made it an explicit condition that the donation stay secret, a friend told Dantzig about it after Koopmans's death. But the professor needn't have had a bad conscience. In the same year that Koopmans was awarded the Nobel, Dantzig received the National Medal of Science. Unfortunately, that honor was not backed up with the kind of cash that is attached to the Nobel, but it's still a nice thing to get. But, the greatest privilege was to have invented a method that continues to be—more than half a century after it was developed—a blessing to humankind. Of course, it also allows military people to wage war more efficiently. "The tremendous power of the simplex method is a constant surprise to me," Dantzig wrote in his reminiscences.

Hales used the simplex method over and over again for the proof of Kepler's conjecture. But he could not simply take a canned program and apply it to his problem. There were two hurdles: his constraints were not always linear and the solutions often had to be integers. Let us turn to the latter hurdle first. We already mentioned (in chapter 4) that it is far more difficult to solve problems for integer numbers than it is to find solutions with real numbers. For illustrative purposes, take the equation A + B = 1. There are an infinite number of real solutions (for example, 0.123456 and 0.876544), but it has just two integer solutions (not counting the negative ones). An airline may serve as a practical example. Its optimal flying stock may consist of 26.7 Boeing 747s and 35.2 Boeing 777s, but it will have great difficulties placing an order for such a fleet. Rounding the numbers up and down to, say, 27 and 35 is no guarantee of optimality either. The optimal integer solution may be far removed from the rounded values.

Hales tackled the integer problem with a technique that had already served him well with the dirty dozen. He looked it straight in the eye and . . . ignored it. Since he was not actually looking for the optimal solution, but just an upper bound, this copout was allowed. The non–integer solution is always at least as good as the more restricted, integer-valued solution—even though it may not be feasible. Then he did a similar thing with the nonlinearities. He replaced them with linear constraints that were less tight than the nonlinear ones. We pointed out in the previous chapter that in optimization problems, as well as in dictatorships, the relaxation of constraints has the effect of raising the optimum. The treatment resembled the placing of a roof over the feasible region. After having made sure that no corner of the feasible region protrudes, the roof's peak can be sought. If it lies below 8 pts then the true, possible noninteger, optimum certainly lies below 8 pts.

A typical example of Hales and Ferguson's problem had between one hundred and two hundred variables, and between one thousand and two thousand constraints. The variables in the linear programs were angles, volumes, and distances. The constraints expressed the conditions on lengths and angles so that only those packings that could actually exist were considered. Nearly one hundred thousand such problems had to be solved in the proof. In 98 percent of the five thousand nets that Hales investigated, that method worked.

But in about one hundred cases the relaxation method did not succeed. By letting the variables take on any value—not just integer values—Hales had relaxed the constraints too much. As a consequence, the roofs' peaks became higher than 8 pts. He would have to try to construct lower roofs. For this a more refined method was needed and he found it in the "branch and bound" method. B&B is an adaptation of linear programming to problems where some or all of the decision variables must be integers.

Say an investment company wants to decide which projects to invest in, given its budget, manpower limitations, and legal constraints. Of course, the company wants to maximize profits. So the computer department runs a linear program and proudly sends the solution in a manila envelope marked "highly confidential" to the CEO, who opens the envelope and starts scanning the results. Some projects carry a *one* (do invest), others a *zero* (don't invest). But then the CEO does a double take: some projects carry fractional results, like 0.716. The CEO stuffs everything back into the envelope and sends it down to the basement—together with a cute little note about nerds.

The nerds scratch their eggheads for a while, and then hit upon an idea. They let the simplex algorithm branch out, trying both zero and one for each project where no clear-cut recommendation had been received in the first run. Both branches are analyzed separately. If the profits for some branch are inferior to previously obtained profits (that is, the profits are *bounded* from above), the branch is ignored. Those branches that are not ignored are further branched. This process continues until all decision variables have become integers. Meekly, the envelope is sent up to the boss again. This time the CEO is satisfied.

In principle, the tree with all its branches could get enormous. After all, there could be twice as many branches at each step as there were at the step before. But generally the bounding process eliminates all but a small fraction of the branches (which may, nevertheless, go into the tens of thousands).

Using B&B, Hales constructed new, tighter fitting roofs above the feasible regions for the one hundred Delaunay stars that had remained. Then he sought the peaks of these new roofs. As he had hoped and expected they lay safely below 8 pts every time. Every time, that is, except for the dirty dozen (see the appendix).

In the course of his work with computers, Hales encountered another obstacle. The problem can be demonstrated with a hand calculator, as long as it has buttons for "square" and "square root." Type in the number 10, hit the "square root" key, and then hit the "square" key. Not surprisingly, the number 10 appears on the display. After all, the square of the square root of 10 is 10. Try again, but this time hit the "square root" button thirty times in a row, and then hit the "square" button thirty times in a row. You would again expect 10 as an answer, but brace yourself. This time the output shows something like 9.5338764. Maybe you should throw your calculator away and use a desktop computer? Go on now, throw the calculator in the dustbin and start up a spreadsheet, like Excel. Enter the number 10 into cell A1 and then enter "=(10^(1/2^A1))^(2^A1)" into cell A2. The answer is 10, as you would expect.[7] Now increase the number in cell A1 to 20.

---

[7] $(10^{1/2^{A1}})^{2^{A1}} = 10$, for any value of A1.

Everything okay? Yes, the answer is still 10. Increase the number to 30. Everything still okay? Yes. Increase the number to 40 and a funny thing happens: The result is 9.999051. Funny maybe, but who cares about 0.000949? Now enter 50 in cell A1 and watch what happens. The displayed answer now is 9.487736 or something similar. Enter 51 and you get 12.18249, nowhere near the correct answer of 10. What happened? Is this a bug? No, it is not a bug, it's a *feature*.

Before we delve into the inner workings of computers to investigate this feature, let us look at two more examples. In the early 1960s a meteorologist at MIT, Edward Lorenz, ran a computer program that simulated weather conditions. At one point he jotted down an intermediate result, and then let the simulation continue on to hailstorms, thunder, and lightning. The next morning he started the evolution of weather again from the intermediate result that he had jotted down the previous day. Then he went to the cafeteria for a cup of coffee. When he came back he had a surprise. Instead of the foul weather that he expected in his virtual world, the sun was shining and a soft breeze was blowing. Why did the computer produce completely different weather conditions on the second run?

The answer is chaos. Chaos is often defined as "sensitivity of the outcome to initial conditions." Lorenz had jotted down the numbers from the first run to three digits after the decimal. He thereby inadvertently truncated the numbers that the computer stored internally to, say, eight digits after the decimal point. By truncating the numbers, he had changed the initial conditions. And since the weather is very sensitive to these initial conditions, sunshine happened instead of hailstorms.

This phenomenon is often called the "butterfly effect," wherein a butterfly flapping its wings in Brazil can produce a thunderstorm in Florida. Even if the wing-flapping corresponds to a disturbance to the air of no more than the twentieth digit after the decimal, the disturbance propagates and expands. By the time it reaches the Sunshine State, the minute disturbance could have jogged itself up to gale force. Incidentally, the butterfly effect also has a good side to it. Since a butterfly in Brazil can disturb the serene weather in Florida, the same butterfly could calm a hurricane in Texas by simply flapping its wings in a certain fashion. This process is called "controlling chaos" and has been put to use with some success in dealing with heart fibrillation. By applying small shocks at precisely the right moment, an erratic heartbeat can be regularized and a heart attack avoided.

The second example is a strange phenomenon that may occur when consecutive measurements are made of a contracting object. This could happen, for example, when a spacecraft that is traveling near the speed of light accelerates. According to Einstein, it becomes shorter at higher speeds. Due to the finite precision of yardsticks, the measurements become less precise the shorter the spaceship becomes. At first this leads to seemingly jumbled-up observations. As the speed increases, these observations organize themselves into a series of interesting dome-shaped patterns. The same patterns, which find their explanation in number theory, can be observed when measurements are performed with increasingly coarse or increasingly fine instruments.

The common thread to the above phenomena is round-off error, or truncation. A yardstick, a calculator, and a computer can measure, compute, and store numerical values only to a certain number of digits after the decimal point. When von Neumann built the first computer he did not yet have to bother with such niceties. But the problem was soon recognized. Different machines used different procedures to round off numbers, and when a program was transferred from one machine to another, strange things happened. $X - X$ was not always equal to zero, and $X - Y$ sometimes was. 0^0 produced an "error" on some machines and 1.0 on others; zero divided by zero occasionally equaled zero, while it produced an "error" at other times; and the sign of zero was sometimes positive, sometimes negative, and sometimes undefined. Blunders and mistakes were inevitable, making machine computation very unreliable.[8]

Computers operate differently from the way we expect mathematics to work. For example, with Carl Louis Ferdinand von Lindemann's proof in 1882 that π is a transcendental number, it was established that this number has infinitely many digits. But for a computer, π is not transcendental at all—it ends after 14 or 28 digits.[9] So how can one in good conscience trust computers that ignore a large part of a number's idiosyncrasy?

It took more than just a little effort to restore confidence. First, there was the tiresome business of the numbers' magnitudes. In the early days computers allocated a fixed width to each number, let's say eight positions before the decimal point and two behind. So when the number 12,345,678.90 was added to the number 0.0123456789 the calculation in effect became 12,345,678.90 plus 0.01. Somehow that seemed grossly

---

[8] Blunders and mistakes were not limited to the early days of electronic computing. One day in late 1994 the computing community woke up to find that when using $X = 4195835$ and $Y = 3145727$, the equation $X - (X/Y)Y$ equals 256 (instead of 0) when run on a computer with Intel Inside! This was the infamous Pentium bug that made it into newspaper headlines.

[9] With great foresight the legislature of the state of Indiana attempted to find a way around this bothersome matter in 1897. The honorable legislators wanted to enforce $\pi = 3.2$ *by law.*

unfair. While the first number was represented exactly, the second carried an error of 23 percent (0.0100000000 versus 0.0123456789). Why should one number be much more precise than another? Just because it is larger?

A method, called floating-point arithmetic, was developed to fight against rampant discrimination. Floating-point arithmetic doesn't care where the decimal point is and always registers the same number of significant digits. After the values of the digits have been established, the decimal point floats in and lands at the appropriate position. So the first number above would be written as 0.123456789 E + 8, the second as 0.123456789 E − 1. The number after the E indicates the landing place of the decimal point: +8 denotes a touchdown eight positions to the right, −1 points to one position to the left.

Floating-point arithmetic was an advance but no panacea. In the mid-1970s anarchy reigned among computer manufacturers. Every company supplied its own mathematical library of elementary functions, which widely varied in quality. The companies also decided on their own where to truncate numbers, how to handle division by zero, what infinity looks like, and so on. As a consequence, peculiarities and eccentricities proliferated. It was a jungle, and the professionals started to worry. It became increasingly obvious that standards were needed to make the different practices compatible.

In 1976 the Intel Corporation began designing a floating-point math coprocessor for its 8086/8 microprocessor. A coprocessor assists the main processor in certain specialized tasks. Initially the executives at Intel were reluctant to launch the project because the marketing people thought there was no market for the 8087 math coprocessor. Then John Palmer, the man in charge of the project, reportedly said, "I'll relinquish my salary, provided you'll write down your number of how many [coprocessors] you expect to sell, then give me a dollar for every one you sell beyond that." The coprocessor project was on its way. (Palmer regretted that the marketing guys didn't take him up on the offer. He could have stopped working and retired a rich man.) One of the first things Palmer did was to recruit William Kahan, a professor of computer science at U.C. Berkeley, as a consultant.

At about the same time, the Institute of Electrical and Electronics Engineers (IEEE) decided that something had to be done about standardization. The IEEE is a very well-respected organization with a century-old history.[10] Under its aegis, meetings were organized in the early 1980s in an

---

[10] It grew out of the American Institute of Electrical Engineers (AIEE), which was founded in 1884, and the Institute of Radio Engineers (IRE), which was formed in 1912. These two organizations formed the IEEE on January 1, 1963.

attempt to reach a consensus on the floating-point issue. Kahan thought it wise for Intel to get in on the standardization efforts and requested permission from the company to participate in the meetings. He was not going to charge the company for the time spent with IEEE, in order to keep his standardization efforts separate from Intel's commercial interests. In the committee he was going to work solely toward the best interest of the community.

Together with a student and a colleague, Jerome Coonen from U.C. Berkeley and Professor Harold Stone, Kahan went to work. A draft was produced, which was to become known as the KCS proposal, after the authors' initials. But the Digital Equipment Company (DEC), arguing that the KCS proposal was too complicated, advocated its own format. Of course DEC had a vested interest in its design, since it was tailor-made to work on its VAX computers. Kahan countered that his proposal was designed not only with experts in floating-point arithmetic in mind, but also programmers not well versed in numerical analysis. So what if the KCS proposal was a bit complicated? At least it managed to satisfy a lot of conflicting requirements.

For years the dispute raged on like a religious war. Kahan was hampered in the meetings of the IEEE committee because he could not disclose too much information. This would have given the competition valuable hints about the design of the 8087 coprocessor. At one point the DEC people argued that a certain feature would be infeasible. Kahan knew full well that this was not true—Intel already had working prototypes—but he could not tell anybody. Another time he proposed very stringent specifications for the "square root" operator. But then he became afraid that the working group would not endorse a standard that was deemed unrealistic. So he lifted the veil just a little and divulged some details about Intel's way of rounding "square root" functions.

Then "underflow" became an issue. The term refers to the question of how small a number must be in order to be flushed to zero with impunity. Kahan proposed a method of "gradual underflow," which wedged a few ultrasmall numbers into the gap between zero and what was previously the smallest number. However, the DEC people expressed fear that gradual underflow would slow their computers down. The company commissioned a highly respected computer scientist to assess the value of gradual underflow. Of course, they expected him to corroborate their claim; after all, they were paying him a fat consultant's fee. To their astonishment, the professor announced that gradual underflow was the right thing to do. This setback put a damper on DEC's enthusiasm for their own design and they discontinued fighting KCS.

Support for the KCS proposal started pouring in from Kahan's former students and, little by little, it became a de facto standard. Finally, in 1985, the IEEE gave its official stamp of approval: henceforth the KCS proposal became known as "IEEE Standard 754 for floating point arithmetic." Many years later Kahan said in an interview that the story of how IEEE 754 developed was "one example . . . where sleaze did not triumph." In recognition of his role in the design of IEEE 754, Kahan was awarded the Turing Award of the Association for Computing Machinery (ACM) in 1989. The ACM's most prestigious award, it is bestowed on an individual who makes a contribution of lasting importance to the computer field. By some it is considered the Nobel Prize in computer science (though it carries a cash prize of only $25,000).

These were some of the issues that Hales and Ferguson had to take into account. The implication of IEEE 754 for Kepler's conjecture was that numbers—small or not—could not be taken at face value. Because of rounding and truncation errors they had to be embedded in intervals. Then, using not the numbers but the intervals, the calculations could be continued. For example, to add the numbers 1.2435826 . . . and 3.5823043 . . . one could produce the *naïve* sum 4.8258869. . . . But this is only *approximately* correct because nobody knows what lies beyond the seventh digit after the decimal point. Now use the intervals [1.2 to 1.3] for the first number, and [3.5 to 3.6] for the second. Add the lower and the upper bounds of the two intervals to form the new interval [4.7 to 4.9]. The correct sum of the two numbers must lie somewhere in the interval. This is not very precise but at least it is absolutely correct. The same procedure can be applied to subtraction, multiplication, and division. With numbers in the intervals $[a,b]$ and $[c,d]$, for example, division would be accomplished by forming the interval $[a/d, b/c]$.[11] Its bounds guarantee that the interval is wide enough to contain the true result.[12] Things are slightly more complicated with square roots and trigonometric function, but the same principle applies.

In real-life applications, round-off errors accumulate as the calculations progress. Consequently, the intervals grow wider and wider. The correct answer remains trapped inside the interval, but as the intervals get larger, care has to be taken that the bounds do not explode into meaninglessness. Hales and Ferguson's way around the explosive issue was to postpone the creation of the intervals to the end of the calculations, rather than forming new intervals at each intermediate step.

Interval arithmetic was the solution to the numerical problems that arose in the course of the proof of Kepler's conjecture. Ferguson summed it up: "[F]loating point arithmetic alone only provides an approximation to the

---

[11] Assuming that a, b, c, and d are positive.
[12] $[a/c, b/d]$ would be too narrow an interval.

correct value of a computation . . . , [it] cannot constitute a proof." But "floating-point interval arithmetic . . . is correct."

With these minutiae out of the way, the question of which programs to use arose. Hales decided on two commercial systems: Cplex to run the linear programs, and Mathematica to do symbolic manipulations. Later Ferguson double-checked the results with Maple.

Cplex is an optimization program that was developed by Robert Bixby, a professor of computational and applied mathematics at Rice University. Bixby had earned his B.S. degree in industrial engineering from Berkeley and his M.S. and Ph.D. degrees from Cornell University. His research interests include linear programming problems in all shapes and forms. Bixby is one of those rare individuals who actually enjoy contemplating numbers. "I find great satisfaction in looking at large volumes of numerical data and deducing new, generally applicable solution strategies," he once related. Together with colleagues from Princeton and Rutgers he broke the world record for the celebrated Travelling Salesman Problem (TSP). This notoriously difficult problem consists of finding an optimal route for a salesman who has to visit a certain number of cities. The route should be the shortest possible, and no city may be visited more than once. In their record-breaking work, Bixby and his colleagues found an optimal path for a TSP with 7,397 cities. They are now working on a TSP with 13,509 cities.

The problem and its solution are not very realistic since it would keep the salesman on the road for about twenty years without ever letting him visit his home. So Bixby and his colleagues also did some more down-to-earth things, like optimize an airline crew scheduling model that included no less than *thirteen million* variables. Such conundrums, plus his teaching duties, kept Bixby busy. But he had some spare time nights and weekends— and spent them designing and writing an optimization program that he would use for illustrative purposes in his classes. He used the computer language C to write the code for his version of the simplex algorithm. So he called the program Cplex.

In 1988, Bixby, Janet Lowe, a business student at Rice, and her husband Todd formed a company around the product. Then the phones started ringing and they've never stopped since. Today, most of the Fortune 1,000 companies are using Cplex. Apart from commercial users, universities throughout the world also employ the Cplex optimizer for linear and integer programming. The University of Michigan was no exception, and Cplex became Hales's program of choice for the one hundred thousand linear programs he had to run. He was so fond of the program, he wrote a little essay for the May 1999 "Ilog Cplex Newsletter," describing how Cplex had helped him solve Kepler's conjecture. (Ilog is the company that now owns Cplex.)

But a program for optimization was not the only software Hales and Ferguson needed. The proof of Kepler's conjecture required handling many formulas and the manipulation of symbolic mathematics (now commonly known as "computer algebra"). What was needed for these tasks was not a number cruncher, like Cplex or Excel, which produces numerical solutions. Rather, Hales sought a symbolic computation system that would express and manipulate complex equations, using automated mathematical formalisms and knowledge systems. He found it in Stephen Wolfram's Mathematica.

Stephen Wolfram was a true child prodigy. Born in London in 1959 to a novelist father and an Oxford philosophy professor mother, Stephen was educated at Eton, Oxford, and the California Institute of Technology. At the tender age of fourteen Stephen neatly typed a book about the physics of subatomic particles. Shortly after he turned fifteen, a scientific paper of his appeared in a learned journal. By the time he reached his twentieth birthday, he had already published a dozen articles, some of which would become classics in their field. In the same year, he received his Ph.D. from Caltech, and at twenty-two he was granted a five-year MacArthur "genius" award.

Armed with the MacArthur money, Wolfram, who had had a fallout with Caltech over patent rights, moved to the IAS at Princeton. At first he continued his studies in cosmology, particle physics, and computer science. But then he discovered an arcane subject that had been overlooked by most colleagues: cellular automata. It was one of Wolfram's predecessors at IAS, the illustrious John von Neumann, who had conceived cellular automata as one of his many sidelines in the late 1940s. But the great mathematician lost interest—his two papers on the subject were published posthumously—and nobody really picked up on the idea.

Until 1970. In that year, John Conway invented the Game of Life. Life is played on a grid on which some black squares are distributed. They could represent, say, bacteria. A few simple rules determine how the situation develops in the next generations. Some bacteria die, some survive, and new ones are born. Then the fun begins. One can watch as generation after generation unfolds and the most amazing things happen. Some initial populations of bacteria simply die out, others persevere in ever changing patterns, and still others just linger for a while and then blink until eternity. There are populations that eat each other up, others that spew each other out, and those that bounce around in a neverending dance. All these phenomena are the consequence of just a few simple rules.

After Martin Gardner published accounts of the Game of Life in his *Scientific American* column, it become very popular among amateur mathematicians. Some scientists also started to take notice, among them Stephen

Wolfram. He took a close look at these strange constructs, analyzed them, classified them, and catalogued their characteristics. The brash young man was convinced that his findings would revolutionize everybody's thinking about science.

But again, few scientists picked up on the idea. It was the time when chaos theory had become the current fad. Every self-respecting physicist, especially those of the younger generation, felt that they needed to have a finger in some chaos-related research. Cellular automata were on nobody's list of favorite topics. In 1986, disappointed that his work did not receive the attention he felt it deserved, Wolfram decided on a career change and became a businessman.

During his scientific research he had developed a piece of software that did scientific computing. Now was the time to make some money off it. Stephen spent the next two years putting his program into commercial form and a revolutionary piece of software was born: Mathematica. The manual that accompanies the system is more than fourteen hundred pages long and weighs over 3 kilograms.

Mathematica can produce numerical solutions to difficult problems, but that is by far the simplest of its features. The fact that Mathematica easily displays π to ten thousand digits, or that it instantaneously computes all 16,325 digits of 5,000! (that's 5,000, multiplied by 4,999, multiplied by 4,998, and so on) are also just minor peculiarities. Finding the prime factors of 1,000,000,000,000,001 is easy $(= 7 \cdot 11 \cdot 13 \cdot 211 \cdot 241 \cdot 2161 \cdot 9091)$ and the billionth prime number is, of course, 22,801,763,489. But Mathematica goes far beyond numerical calculations. It does symbolic mathematics. That means that when you want to transform, convert, modify, change, and transmogrify a formula, Mathematica will give you the desired result in symbolic form. For example you can ask for the factorization of $(x^{25} + y^{25})$ and get the answer as a formula.[13] You can enter a function that you want to integrate, and out comes the correct expression. And if it does not, you may be reasonably sure that a solution to your question simply does not exist. Then there also are Mathematica's truly amazing graphics. They have become the standard against which competitors are measured. Mathematica even allows you to visualize functions in four dimensions, as movies.

In the following years, Wolfram showed his mettle as a businessman. He built his company, Wolfram Research, into a miniconglomerate with three hundred employees. T-shirts, posters, mugs, and baseball caps sporting the company's logo are an integral part of the company's marketing strategy, as

---

[13] $x^{25} + y^{25} = (x + y)(x^4 - x^3y + x^2y^2 - xy^3 + y^4)(x^{20} - x^{15}y^5 + x^{10}y^{10} - x^5y^{15} + y^{20})$.

is a journal devoted entirely to the system, and a Mathematica-Mobile that travels the world preaching the gospel according to Stephen Wolfram. An estimated two million engineers, industrialists, scientists, and students use Mathematica all over the world.

So Hales opted for Mathematica to do his symbolic mathematics. Not so Sam Ferguson. Tom and Sam had decided early on that since their proof relied so heavily on computer calculations, they would have to double-check their programs. They independently rewrote large parts of the programs using different software so that possible shortcomings of the programs could be excluded as a source of error. Sam's program choice was Maple.

At a time when Wolfram was still developing Mathematica, Maple was already in its version 4.2. Maple was developed at the University of Waterloo in Ontario, Canada, by Professor Keith Geddes, codirector of the Symbolic Computation Group of the university, and Professor Gaston Gonnet, now head of the Institute for Scientific Computing at the ETH in Zürich. The two professors were interested in programs that could perform symbolic mathematics.

Research work on the design of systems for performing computer algebra had been underway since the 1960s, long before either Wolfram or Geddes and Gonnet entered the picture. But the early systems, which were developed during the 1970s, were very large, demanded many megabytes of RAM, and required extensive computer time to perform routine mathematical computations. Consequently, only a tiny number of researchers with access to large mainframe computers were able to exploit this technology. That's where Mathematica and Maple came in.

Geddes and Gonnet began their collaboration on the project in November 1980. Their primary goal was to design a computer algebra system that would be accessible to researchers in mathematics, engineering, and science, and to large numbers of students for educational purposes. Only three weeks after they started work, they already had a functioning prototype. Within a couple of months the system was being used at the University of Waterloo to support courses for graduate and senior undergraduate students. At the end of 1983 the software was installed on about fifty external computers. By 1987 there were about three hundred installations and the professors decided to go global. Waterloo Maple Software (now Waterloo Maple, Inc.) was incorporated in 1988, the same year Stephen formed Wolfram Research, Inc.

The Canadians were more down-to-earth than their extravagant competitors south of the border. Neither baseball caps nor umbrellas were distributed. No T-shirts bore the maple leaf and no Maple-Mobile spouted the gospel. Maple was a plain vanilla version of the flamboyant Mathematica. As a consequence, the installed base grew modestly to about two thousand installations during the next two years. But starting in 1990, increased attention was paid to sales and marketing. Even though there were still no baseball caps, estimates at the start of the twenty-first century run to about one million users worldwide.[14]

---

[14] Gonnet, who remains a major shareholder of Waterloo Maple, has had a rocky relationship with the company he helped found. The mother company sued him over ownership of Maple's technology, and Gonnet did not hesitate to countersue.

CHAPTER 13

But Is It Really a Proof?

Sunday morning, August 9, 1998. The previous day marked the ninety-eight-year anniversary of David Hilbert's famous speech at the Second International Congress of Mathematicians in Paris. Tom Hales was finally done. He sat down to write a message announcing that Kepler's conjecture was no longer a conjecture. At five minutes to ten he sent out the e-mail to his colleagues around the world.

From hales@math.lsa.umich.edu
Date: Sun, 9 Aug 1998 09:54:56 -0400 (EDT)
From: Tom Hales

Subject: Kepler conjecture

Dear colleagues,

I have started to distribute copies of a series of papers giving a solution to the Kepler conjecture, the oldest problem in discrete geometry. These results are still preliminary in the sense that they have not been refereed and have not even been submitted for publication, but the proofs are to the best of my knowledge correct and complete.

Nearly four hundred years ago, Kepler asserted that no packing of congruent spheres can have a density greater than the density of the face-centered cubic packing. This assertion has come to be known as the Kepler conjecture. In 1900, Hilbert included the Kepler conjecture in his famous list of mathematical problems.

In a paper published last year in the journal "Discrete and Computational Geometry" (DCG), I published a detailed plan describing how the Kepler conjecture might be proved. This approach differs significantly from earlier approaches to this problem by making extensive use of computers. (L. Fejes-Tóth was the first to suggest the use of computers.) The proof relies extensively on methods from the theory of global optimization, linear programming, and interval arithmetic.

The full proof appears in a series of papers totaling well over 250 pages. The computer files containing the computer code and data files for combinatorics, interval arithmetic, and linear programs require over 3 gigabytes of space for storage.

Samuel P. Ferguson, who finished his Ph.D. last year at the University of Michigan under my direction, has contributed significantly to this project.

The papers containing the proof are:

An Overview of the Kepler Conjecture, Thomas C. Hales
A Formulation of the Kepler Conjecture, Samuel P. Ferguson and Thomas C. Hales
Sphere Packings I, Thomas C. Hales (published in DCG, 1997)
Sphere Packings II, Thomas C. Hales (published in DCG, 1997)
Sphere Packings III, Thomas C. Hales
Sphere Packings IV, Thomas C. Hales
Sphere Packings V, Samuel P. Ferguson
The Kepler Conjecture (Sphere Packings VI), Thomas C. Hales

Postscript versions of the papers and more information about this project can be found at http://www.math.lsa.umich.edu/~hales

Tom Hales

samf@math.lsa.umich.edu
hales@math.lsa.umich.edu

The minute the e-mail was on its way, Hales felt as if a load had been taken off his shoulders. His gamble had paid off. When he had started his work on the proof, and even much later, it had by no means been certain that his efforts would eventually be crowned by success. There had always been a real possibility that his ideas—finding the maximum score of hybrid stars—would lead nowhere. But at long last, everything was worked out and all the skeptics who had expressed doubt about Tom's approach had been proven wrong. Five years of hard labor had come to an end.[1]

---

[1] But maybe Hales should have waited another two years with his announcement. On May 24, 2000, the Clay Mathematics Institute of Cambridge, founded by the Boston businessman Landon T. Clay, announced seven prizes of $1 million each for the solution of open problems. "The Scientific Advisory Board of CMI selected these problems, focusing on important classic questions that have resisted solution over the years." Had Hales waited, the advisory board could have chosen Kepler's conjecture as one of the prize questions, and Hales and Ferguson could have laughed all the way to the bank.

Four weeks later, Hales was on a flight to Israel. He was going to give his first lecture on the completed proof of Kepler's conjecture. The occasion was the fourth triannual congress of the International Society for the Interdisciplinary Study of Symmetry. ISIS-Symmetry is an organization that includes all disciplines and comprises all fields that have anything to do with symmetry: mathematics, of course, but also biology, physics, chemistry, crystallography. And psychology, neurology, and linguistics. And also architecture, choreography, music, and art. The conferences are truly interdisciplinary. Specialists from different areas of study do not just talk past each other but actually communicate. Crystallographers roam around the arts exhibit, architects listen to botanists, biologists talk to theatre impresarios, physicists watch films about modern ballet. People from different fields wander into various lectures, conferences, and multimedia presentations about disciplines they never even knew existed.

The most appropriate venue for a public announcement of Hales's success would have been the Twenty-third International Congress of Mathematicians in Berlin, which had taken place three weeks earlier. Nearly thirty-five hundred mathematicians from one hundred countries participated in this mammoth event that is now organized once every four years by the International Mathematical Union.[2] Unfortunately, the deadline for the registration of lectures had been May 1.[3] At that time, in the late spring, Hales had not yet been sure whether his proof would be completed soon enough for the congress. Then, during the summer, progress was faster than he expected, but it was too late to announce a presentation. *Ordnung muss sein!*—especially in Berlin.[4] At least Hales had the satisfaction that in the wake of his e-mail his feat was already known by most participants at the congress.

ISIS-Symmetry was holding its congress in Haifa on the Mediterranean. The city in the north of Israel was a particularly well-chosen location for this conference. Haifa is the seat of the holiest shrine of the Baha'i faith, and the gardens that surround it are world-famous for their beauty and symmetry. In his briefcase Hales had brought slides for a presentation with him to Haifa. The organizers of ISIS-Symmetry made no fuss about any missed deadlines. This was the Middle East, after all, and people were more relaxed

---

[2] There were no congresses during the First and the Second World Wars.
[3] It was exactly this deadline that David Hilbert had missed ninety-eight years earlier.
[4] *"There must be order!"* This Teutonic utterance has been used by civil servants and bureaucrats throughout the ages. It justifies the call for compliance with even the silliest requirement. It is an axiom with which one simply does not argue. To be fair, the organizers of the Berlin conference were not quite as thick headed as that. There was a provision at the congress for "those who have the wish to make a spontaneous contribution."

about dates and regulations. *Yallah,* forget about the deadline.[5] Everybody was more than happy to have Hales present his proof. The congress began on September 13, 1998. At the opening session, the chairman of ISIS-Symmetry announced the presence of the man who had proved Kepler's conjecture, four centuries after it had first been proposed. The audience spontaneously rose to their feet and gave Hales a standing ovation.

On the fourth day of the conference, he delivered his talk. Again there was wild applause. But there was no scrutiny of his work at all in Haifa. The listeners—who included the architects, the choreographers, and at least one journalist—were delighted to witness such a momentous event. But they did not ask any incisive questions. And the physicists who attended Hales's lecture had, of course, known all along that Kepler's conjecture was true.

The first real test came four months later, at the beginning of 1999. The e-mail that Hales sent out in August had also reached Robert MacPherson at the Institute of Advanced Studies. He was intrigued and organized a mini-conference at IAS, entitled "Workshop on Discrete Geometry and the Kepler Problem." The announcement spelled out the theme of the workshop: "The main subject of the workshop will be the new computer-assisted proof of the Kepler Conjecture by Tom Hales. Most of the talks and working sessions will be . . . on this subject." The flyer went on to explain to those who might not know that "the Kepler Conjecture asserts that the densest packing . . . is the one we all expect."

The workshop was scheduled for a full week and luminaries on sphere packings and discrete geometry were to attend. Gabor Fejes-Tóth, Laszlo's son, was appointed the organizer. Hales was the informal guest of honor. Sam Ferguson was invited and came. So did John Conway and Neil Sloane. And of course Wu-Yi Hsiang also attended. Altogether about a dozen people from as far away as Hungary and Austria assembled in Princeton from January 17 to 22. The major lectures took place every day at 2 P.M., with shorter talks scheduled in between. Ferguson gave the opening address on Sunday and Hsiang's talk was scheduled for Tuesday.

The workshop was very intense and many tough questions were asked. But Hales never wavered. His answers were clear and to the point. This reassured the audience. Even Hsiang did not have any technical objections. His only critique was to the effect that the proof was not very pretty. Hales never contested that. "We all agree that this proof is ugly," Ferguson commented later. "However, it was the best that we could do, at least up to that

---

[5] *"Go ahead already."* This Arabic/Hebrew expression is the Middle Eastern counterpart—and exact opposite—of *Ordnung muss sein!* It justifies just about every infraction.

point. Hales could spend the rest of his career simplifying the proof. But that doesn't seem like an appropriate use of his time." Initially, there had been some apprehension about how Hsiang would react to all the hoopla about Hales's proof. But he remained quite friendly, if somewhat condescending. After Hales's talk he commented, "I am glad that Kepler's conjecture has now been proven once again." His perspective seemed to be that everything is much simpler if you apply the appropriate geometric insight. Then he gave a one-hour presentation himself, entitled "On the proof of Kepler's conjecture." Of course he meant *his* proof, not Hales's. But he did not manage to convince anyone.

The participants came away from the meeting convinced that Hales had made his point. The fine print would still have to be checked, of course, but overall the proof seemed to hold up. After the workshop was over, MacPherson thought long and hard about Hales's achievement. He was one of the six editors of the *Annals of Mathematics,* one of the most respected, if not *the* most respected mathematics journal in the world. Would Hales's proof be a worthy candidate for publication in the *Annals?*

The *Annals of Mathematics* was founded in 1884 by a professor at the University of Virginia, Ormond Stone, who financed the journal out of his own pocket for the first ten years of its life. It is the second oldest mathematics journal in the United States.[6] In 1899 the editorial office moved to Harvard, and in 1911 to Princeton, where it is still located today. Starting in 1933 the IAS got a foot in, and since then *Annals* has been edited jointly by the mathematics department of Princeton University and the IAS. At the beginning of the twentieth century, the subscription cost $2 a year, and most mathematicians in the United States subscribed to it. After the First World War the *Annals,* which had hitherto mainly published papers from American authors, gained in importance. European journals were undergoing difficult times and many mathematicians from abroad started sending their papers to the *Annals.* In response to the increased influx during the 1930s, the *Annals* expanded its publication schedule from four issues a year to six. In the late 1920s, Solomon Lefschetz took over as editor, a position he kept until 1958.

Lefschetz was an algebraic topologist—in fact, the word "topology" came into use only after Lefschetz wrote a pathbreaking monograph with this title in 1930. However, his claim to fame did not only derive from his outstanding mathematical abilities, but also from the fact that he had two artificial

---

[6] The oldest mathematics journal in the United States is the *American Journal of Mathematics,* whose editorial offices are located at Johns Hopkins.

hands. He lost the limbs in a laboratory accident when he was twenty-three years old. As a result, he had to turn away from his original calling as an engineer and became a mathematician. But the engineers' loss turned out to be the mathematicians' gain, as Lefshetz became one of the most influential mathematicians of his time. The door to his office at Princeton had a special hook installed, instead of a knob, so that he could open and close it with his lower arm. And instead of putting the standard issue filing cabinets into his room, he received a specially designed set of drawers that enabled him to store and find papers with relative ease. Every morning an assistant would put a chalk into his shiny, black prosthesis and Lefschetz would jot down equations on the blackboard in enormous letters, like a child learning how to write. In the evening an assistant removed the chalk.

During his thirty-year tenure, Lefschetz put a lasting mark on the *Annals.* The only other publication in the 1930s that came close to it in terms of the quality of published papers was the *Transactions of the American Mathematical Society.* However, that journal was not to everybody's liking. The board was extremely fussy, refereeing was heavy handed, and publication was slow. The editors did not like short papers. Neither did they like long papers. The *Annals,* on the other hand, was game for anything between two pages and one hundred pages. And refereeing was not always an impediment to publication. "[Lefschetz] would hear of some new result, . . . solicit the paper and promise publication without refereeing," a colleague related. This could have been disastrous, except that Lefschetz had an uncanny instinct for good and important work. Whenever a refereeing job was called for, it was unusually fast, because most of it was done in-house, at Princeton. One of the referees was John von Neumann, Lefschetz's coeditor. It is told that he refereed even the most difficult papers just by flipping through the pages. That did not mean that he didn't read the paper; he did. It was just that his brain worked ten times faster than anybody else's.

Lefshetz's autocratic style as editor of the *Annals* had its disadvantages. "[He] made a lot of enemies, because two people, say, would be competing to get first publication in some new thing and the person who published in the . . . *Transactions* got nipped by the guy who got it in with Lefschetz." From a fairness standpoint, the *Transactions*'s editorial policy was by far the more ethical. "The *Transactions* was run democratically, no favors, everybody treated the same. The *Annals* was run with a great deal of favoritism." But favoritism is not always considered a drawback, especially by those on the receiving end. And for Princetonians, publication in the *Annals* used to be de rigeur anyway.

Lefshetz ruffled some other feathers too. When he was about to be nominated to the presidency of the American Mathematical Society, the eminent mathematician George D. Birkhoff at Harvard wrote to a friend: "I have a feeling that Lefshetz will . . . try to work strongly and positively for his own race. They are exceedingly confident of their own power and influence in the good old USA." Take a quick guess what Lefshetz's religious background was. Birkhoff's letter went on: "He will get very cocky, very racial and use the Annals as a . . . racial perquisite. The racial interests will get deeper, as Einstein's . . . do." Later on, Birkhoff did redeem himself a little, as one Jewish contemporary testified: "In all fairness it should be noted that, in spite of his stated position on refugees, Birkhoff did help some [Jewish] refugees get positions in less prestigious schools." The anti-Semitic attitudes that prevailed among the American intelligentsia in the 1930s led even Lefshetz to propose a quota for Jews. He once remarked that no more Jewish graduate students should be admitted to Princeton, because Jews could not get a job anyway, so why bother.

Today, the *Annals* remains one of the premier outlets for mathematicians. This declaration is no empty statement; it can be made precise. Customarily, a journal's impact on science is measured by the number of times its articles are quoted by other authors. By that count the *Annals* comes out close to the top. Its articles are cited, on average, 1.71 times each year in the two years following publication. This puts the *Annals* in third place in terms of impact after the *Bulletin of the American Mathematical Society* (1.88) and *Computational Geometry* (1.82). The *Transactions,* with an impact factor of 0.55, ranks about twenty places lower.

In order to safeguard the *Annals*'s reputation as the world's most prestigious mathematics journal, every article that is submitted must undergo a very stringent refereeing procedure. This is hard, even on the editors. "I don't enjoy rejecting nine out of ten papers," MacPherson has remarked. But he would have certainly liked to see the paper containing the proof of Kepler's conjecture published in the pages of the *Annals,* which was easier said than done. After all, MacPherson was not Lefschetz and had no desire to emulate his editorial policies. So he first polled his coeditors. He sent e-mails to his five colleagues on the editorial board asking for their opinions. He cautioned them that the paper would be very long. He also noted that Hales's proof was computer-aided. Under previous editors the *Annals* would never have accepted anything that contained something as ignominious as a computer proof. But times have changed and MacPherson's coeditors were game. They unanimously recommended that he solicit the paper.

Hales was asked whether he would be interested in publishing his work in the *Annals.* MacPherson told him right away that it would be a very drawn-out process. Nothing but an extremely rigorous refereeing process would be acceptable. As an example he cited a similar paper, also a computer-assisted proof, whose audit took two teams of referees three full years. Hales did not mind. He himself wished for a thorough and critical reading of his work. He did not want any doubts lingering on for years, and would welcome a stamp of approval.

But MacPherson had another fear. The work that he had seen so far was not particularly well written. During the years that Hales had spent working on the various parts of his proof, he had recorded everything in a very meticulous fashion, as in a lab report. The moment the computer was done, the appropriate manuscript was also done. As a result, the papers were not easy to read. And since they were completed at different times, various parts had to be reformulated and retrofitted. Accordingly, the proof resembled a patchwork of loosely related bits and pieces. Would Hales agree to revise his work? Ferguson certainly was not very enthusiastic about the prospect. "The editors of the *Annals* want a comprehensive, up-to-date presentation, hopefully simplified. All well and good, in principle, but both of us are tired of working on the problem, which we consider to be solved, and properly documented. It seems a bit much to ask us to re-work the entire proof merely to make the proof more accessible."

So how does one go about refereeing a proof that consists in a large part of computer programs? MacPherson was not overly worried. "When I referee a paper I try to understand the internal logic of the proof and do consistency checks. I don't check the proof statement by statement." An audit of a computer-assisted proof would follow the same guidelines. For the proof of Kepler's conjecture twelve referees were chosen, mostly from among the workshop participants, and Gabor Fejes-Tóth, who has a reputation of being a careful and responsible organizer, was asked to coordinate the process. As of 2002, the referees, whose names were not revealed to anyone, continue to be very dedicated and most have kept up their enthusiasm. The selfless work of these referees who labor away at someone else's work instead of furthering their own careers is a constant source of astonishment to MacPherson: "Isn't it amazing? They do it for the mathematical community." In Hungary a group of professors and graduate students are running seminars about the proof. Laszlo Fejes-Tóth, Gabor's father, still takes a keen interest in what is happening. Nevertheless, for some referees the work has become too much and they have dropped out of the refereeing process.

But while the proof of Kepler's conjecture was on its way to becoming an accepted part of the mathematicians' bag of wisdom, questions started to arise: Is it really a proof? Can a mathematical truth be demonstrated by brute force? Could a computer be wrong? How do computer-aided proofs compare with the elegance of conventional proofs? What have we learned from the proof?

The first time a computer was utilized not just to solve numerical problems, but to prove a theorem, the mathematics community was up in arms. It was in connection with the so-called Four-Color Problem. The problem had been raised in 1852 by Francis Guthrie, a student of University College, London. He had been given the task of coloring a map of the counties of England. No two adjacent counties were to be painted in the same color, although counties of the same color were allowed to share a corner. After a little while he noticed that he did not need more than four colors. In a letter to his younger brother Frederick, he raised the question whether this was true for any map. Frederick did not know the answer either. Neither did his teacher, the celebrated mathematician Augustus de Morgan. De Morgan wrote a letter to his equally famous colleague, Sir William Hamilton, who also didn't have an answer. For more than 120 years many people tried their hand at the problem but invariably their efforts remained without success. Finally, in 1976, the German-born Wolfgang Haken and his colleague at the University of Illinois, Kenneth Appel, astounded the public with a proof. Overnight the Four-Color Problem had become the Four-Color *Theorem*.

But the mathematical ingredients were provided—horror of horrors—by a computer. Appel and Haken had established 1936 different map configurations that could be possible counterexamples to the conjecture. Then the computer checked each and every one of these prototypes. The computer churned on for about twelve hundred hours. Finally the two mathematicians were able to cry: *Eureka!* Or rather, *Ouk Eureka!* (Didn't find it!) Not a single instance was discovered among the 1936 maps that would have required five or more colors. With that, the Four-Color Problem was proved. Does Appel and Haken's strategy seem vaguely familiar? It does, indeed. It bears an eerie resemblance to Tom and Sam's proof. In fact, the strategy—reduce a proof to a finite list of possible counterexamples, and then eliminate them one by one—is now a mainstay of computer-aided proofs.[7]

Computer programs exist that are perfectly able to discover truths. Genetic algorithms, for example, have discovered laws of nature—without proving them, however, or explaining why they are true. Genetic algorithms are programs that evolve according to Darwin's laws. Such an algorithm starts out with bits and pieces of possible solutions to a problem, and uses them as building blocks for future generations. The building blocks combine, split, and recombine to produce offspring. These offspring become

---

[7] The other computer proof published in the *Annals* (referred to by MacPherson) also followed a similar strategy.

better the longer the computer runs. After many generations, which may take no more than a couple of minutes on a fast computer, the algorithm may produce a formula that mimics the phenomenon that underlies the data.

For example, feed the computer data from the orbits of the planets, and let the genetic algorithm run for a few dozen generations. It invariably discovers Kepler's third law on planetary motion. That may not be totally surprising, since astronomers have known the true answer since the seventeenth century. But take sunspots, for example, a periodic phenomenon of the sun about which we still do not know very much. Feed a century's worth of sunspot data into a genetic algorithm, and out comes a formula that predicts future sunspot activity to a surprising degree of accuracy. Then there are computer programs that perform "data mining." They sift through enormous databases and find connections that nobody would ever have been able to spot. For example, so-called neural network programs determine the buying patterns of legitimate credit card holders and discover fraud as it happens. None of the above examples constitutes a proof. However, such programs tell us where to look for the answer . . . or for the con artist.

Since Appel and Haken first presented their work to a packed audience in Toronto in 1976, it has become a landmark proof in the history of mathematics. But at the time, most mathematicians were horrified. The philosopher Thomas Tymoczko wrote, "If we accept the four color theorem as a theorem, then we are committed to changing the sense of 'theorem,' or more to the point, to changing the sense of the underlying concept of proof," and another purist reserved the right "to reject any proof that emanates only from a 'black box' . . . with the same vigor that I reject a Jehovah's Witness's proof." How can one believe a proof if one cannot verify its every step? Tymoczko rejects Appel and Haken's demonstration because "no mathematician has seen a proof of the four color theorem" and "it is very unlikely that any mathematician will ever see a proof."[8]

We generally don't trust what we can't see. Or do we? Remember Gene Hackman's remark to Denzel Washington in *Crimson Tide,* just before their submarine started its dive? "Last breath of polluted air for the next 65 days. Gonna miss it. I don't trust air I can't see." It's funny because it's so absurd. Of course we trust air, especially if we can't see it. So do we have to watch as a computer processes bits and bytes through its CPU, better yet through its innumerable transistors, before we trust it? Or do *pure* mathematicians

---

[8] Thomas Tymoczko was educated at Harvard and Oxford and became professor of philosophy at Smith College in 1971. He died of stomach cancer in 1996 at the age of fifty-three.

have some irrational prejudice against innovation, like some religious fundamentalists have against medical progress? There is evidence that this may be so. As one mathematician at Harvard found to his surprise, half of his colleagues in the department did not even know how to program. And when an inventory of computing equipment was made at Stanford University, it turned out that the mathematics department had fewer computers than even the French department. The prevailing feeling among purists was reflected by the opening words of Tymoczko's essay "Computers and mathematical practice": "Computers have been intruding upon mathematics for several decades."

But wariness with respect to computers is not just based on prejudice. After all, bloopers do happen. Generally, one distinguishes between two types of errors: human errors (input or programming) and system errors (software or hardware). Human errors—which are known by the term GIGO (garbage in, garbage out)—should be brought to light through a scrupulous refereeing process. But system errors could go undetected even in the wake of the most painstaking audit. Since the introduction of IEEE Standard 754, the rounding of numbers no longer presents a source of error. But the possibility exists that errors are introduced by defective computer chips (as was the case with the Pentium bug) or by faults in the way a computer translates a program into instructions to a microprocessor (compiler errors). Worse still, even a perfectly healthy computer is not totally error-free. A Cray-1A supercomputer, for example, was reported to produce approximately one undetected error per thousand hours of operation. Usually this occurs through a random change of a *bit* in computer memory, engendered by, of all things, cosmic radiation. Manufacturers use error-correcting memory to minimize such faults, but the problem cannot be completely eliminated.

Pierre Deligne, an algebraic geometer at the IAS, is convinced that the human mind is still the measure of all things. "I don't believe in a proof done by a computer," he says, "I believe in a proof if I understand it." Doron Zeilberger from Temple University finds himself at the other end of the spectrum. He proposes a new semirigorous mathematical culture in which computers will be used to establish not the truth, but only the probability of a truth. This would produce statements like "The Goldbach conjecture is true with a probability of 0.9999 and its complete truth could be determined with a budget of $10 billion."

The future could even see the validity of a proposition established by comparing it with experiments run on computers. David Epstein from the University of Warwick in England established a journal devoted to results and conjectures suggested by experiments. For reasons such as these, the mathematician Edward Swart suggested the creation of a new term: *agnogram*.

An agnogram would be a statement that lies somewhere between a conjecture and a theorem. Its veracity would have been verified as well as possible, but its truth would not be known with the kind of assurance one attaches to theorems. About agnograms one would thus remain, to some extent, agnostic.

There is another complaint against brute-force computer proofs. A good conventional proof does not just tell the mathematician that a fact is true, but also why it is true. It allows a deeper understanding of the inner structure of a mathematical system, opening avenues for further discoveries. A computer proof reveals nothing, apart form the truth of the fact. So, after studying the proof of Kepler's conjecture, one has to ask the question: Did one learn anything from the proof? Did one gain a deeper insight into mathematics? Has one become wiser after studying the proof? The painful answers are clearly in the negative. After all, everybody knew that Kepler's conjecture was true, even before work on the proof started. So there was nothing to be gained on that account. Furthermore, the proof did not open any new avenues.

After Hales announced his proof, complaints started pouring in. *No, not brute force again!* Quibblers liked to point to Andrew Wiles's gigantic proof of Fermat's theorem (published, in 1995, in the *Annals of Mathematics*). Now *that* was a classy proof. It was beautiful, it had elegance, it had style. It was in a totally different league. Ian Stewart, the well-known English mathematician and popularizer of mathematics, remarked very aptly that while Wiles's proof of Fermat's theorem resembles *War and Peace,* Tom's proof of Kepler's conjecture resembles a telephone directory. Short it is not. Elegant it is not. Aesthetic—only if you have a penchant for telephone books.

Nevertheless, since Appel and Haken's trailblazing work, the profession has moved along. Computer proofs are, if not universally liked, at least widely accepted as a necessary evil by many modern mathematicians. Hales's own attitude towards computers has changed over time. "I used computers as part of the 1998 proof only because I could not conceive of a way to prove the Kepler conjecture without them. [Lately] my stance shifted considerably, and I now feel that computer proofs are vital to the progress of mathematics."

A protocol of sorts has emerged to minimize the probability of errors. Check the results by hand whenever possible. Solve the same problem using different programs. Check for internal consistency of the results. Don't reinvent the wheel by writing your own programs—leave that to the specialists. But avoid freeware. Instead, use popular, well-known software packages that have stood the test of time. Run the programs on more than one machine, with different processors and with different compilers.

Finally, try to get other people to perform independent verification of your proof. None of the proposed measures can guarantee the absence of mistakes, but they can serve to increase confidence in computer-aided proofs.

Getting other people to perform verifications of computer-based proofs is the hardest problem. In chemistry or microbiology it is quite customary to have graduate students all over the world run experiments dozens, if not hundreds of times. Maybe this should become the norm in mathematics too. Ferguson would clearly like to see independent verification: ". . . it would make more sense for a third party to independently verify our work, rather than asking us to just simplify it. . . . It [doesn't seem] that the mathematical community is willing to ante up the appropriate investment."

Apart from independent verification, Hales and Ferguson complied with most of the suggestions. They used well-known software programs (Cplex, Mathematica, and Maple), and independently wrote key portions of the programs twice. Ferguson also ran parts of the programs at home on his Macintosh PowerPC, to check the results against those obtained from the Sun workstation at the university. "I looked at the output of the compiler on my home machine, using a disassembler, and verified that no errors seemed to be occurring." But the pitfalls remained. "It is difficult to be sure. . . . We tried to be as careful as possible, but we are human, after all." And this, in a nutshell, may be the best reason yet to use computers for mathematical proofs. It is not the fallibility of computers that should be the issue, but human frailty. After all, a computer error of one bit every one thousand hours of operation compares quite favorably with the human brain's rate of bloopers. Thus the adversaries of computer proofs could be beaten at their own game.

When MacPherson initially polled his colleagues, all agreed that computer-aided proofs of mathematical theorems should not be held up to higher standards of rigor than traditional proofs. The mathematical community must not cut itself off from such a resource. After all, even calculus was looked at with suspicion in the late seventeenth century, shortly after it was invented. Besides, errors can also occur in a conventional proof. Shortcomings should become apparent during the refereeing process. But the system does not always work that way. Sometimes an error in a conventional proof appears only after the paper has appeared in print, sometimes a long time after publication. For example, Alfred Bray Kempe, a London barrister and specialist in ecclesiastical law, published the first purported proof of the Four-Color Theorem in 1879 in the *American Journal of Mathematics*. It was considered correct until 1890, when Percy John Heawood, a lecturer at Durham in England, showed that the proof was flawed. (That did not prevent Kempe from receiving great praise in the meantime, from being elected to the Royal Society and from receiving a knighthood.) Bender's faulty proof of the kissing problem, and Hoppe's equally faulty correction are further examples. So is Hsiang's alleged conventional proof of Kepler's conjecture.[9]

We never know if a well-established proof will turn out to be deficient one day. Only errors can be established conclusively; correctness can only be assumed as long as no countervailing evidence appears. After a while, most mathematicians start to *believe* that a proof is correct, simply because nobody found anything wrong with it. Doesn't belief run completely against everything we know about mathematics and mathematicians? I still remember my mathematics teacher in seventh grade telling us: "Don't believe anything I say, unless I can prove it." That sounded very liberal, especially since it came from an authoritarian pedagogue at a junior high school in Switzerland, where discipline was rated higher than intelligence (which, in turn, was rated lower than athletic ability). Was he wrong?

Mathematics is a social process. Truths get accepted by consensus or, lacking that, through the approval by the majority or, lacking even that, through endorsement by the few qualified specialists. Until the beginning of the last century, proofs were short and could be surveyed in one sitting. This has changed. Nowadays there is a lot of trust involved, since not everyone can work his or her way through a hundred-page paper. Only very few people have actually read Wiles's proof of Fermat's theorem. But laypeople trust the judgement of mathematicians, mathematicians trust number theorists, and number theorists trust the verdict of the referees. In the end there may be no more than a few dozen people who actually read and understood a proof. But the whole world *knows* that the theorem is true. As was pointed out in chapter 6, the classification of finite, simple groups involves about five hundred separate articles, written by about a hundred mathematicians, and covering approximately fifteen thousand pages. But there was only one mathematician, Daniel Gorenstein, who had a grasp of the whole project. Since his death in 1992 there is probably nobody on earth who can personally vouch for the correctness of the classification. Seen from that perspective, a proof provided by an impartial computer may be more believable than a conventional proof fraught with potential traps.

[9] See chapters 6 and 9.

CHAPTER 14

Beehives Again

We have nearly reached the end of the saga. Kepler's conjecture was solved and Tom Hales could lie back and bask in the glory of his proof of an age-old problem. But a mathematician's work never ends. The successful solution of one problem opens up new avenues, engenders further conjectures, and spawns novel theories.

The first of the new avenues involved one of Hales's students at Michigan and concerned another long-standing conjecture. Remember the dodecahedral conjecture that Laszlo Fejes-Tóth formulated in 1943? It deals with the configuration of twelve spheres arranged around a central nucleus in such a way that one lies on top, five are arranged in a regular fashion a bit above the equator, five more a bit below the equator, and one on the bottom. This configuration, whose Voronoi cell is a dodecahedron, fills 75.46 percent of space. Thus it is denser than Kepler's arrangements, but since the dodecahedron cannot fill space without gaps, it is useless as a global packing. However, Fejes-Tóth firmly believed that no local arrangement could be denser than that. At first, he thought he had found a proof of this assertion. But then the proof turned out to be faulty. Ever since, his statement was considered a conjecture. On the one hand, no counterexample to the conjecture could be found. On the other hand, nobody was able to prove it either. For a brief period in 1993, Wu-Yi Hsiang thought that he had proved both Kepler's conjecture and the dodecahedral conjecture in one fell swoop, but then gaps showed up in Hsiang's paper, and henceforth the proof was ignored by the mathematics community.[1]

Enter Sean McLaughlin, an undergraduate clarinet student at the University of Michigan. Sean had performed successfully with the Detroit and Toledo symphony orchestras, and came to Ann Arbor to study music. But

---

[1] See chapter 10.

he did not limit his activities to playing the clarinet. The program distributed at one of his performances stated that "in addition to his musical studies [Sean] pursues an interest in mathematics." That was rather an understatement because mathematics was more than just a hobby to McLaughlin. In fact, he took his mathematical studies extremely seriously.

One of his teachers was Tom Hales, who showed him the power of brute force. A sensitive musician, McLaughlin preferred the beauty and elegance of traditional mathematics and he took to computer proofs as he would have taken to freestyle wrestling. Nevertheless, when Hales suggested that the two of them use brute force to attack the dodecahedral conjecture, McLaughlin agreed. That was in the summer of 1997, when Sam Ferguson was in the midst of his thesis work on the dirty dozen. The by-now very experienced Ph.D. student spent many hours discussing and explaining the methods and techniques to his younger colleague.

McLaughlin started working along the lines that Hales had described to him. He began by establishing eight properties that stars and nets must have in order to be potential counterexamples to the conjecture. Then he identified all cases that fulfill these eight requirements. There were about a thousand. Then McLaughlin started his offensive, using linear optimization and interval arithmetic. In all but thirteen cases he succeeded in eliminating the potential counterexamples. Applying even more brute force, the thirteen exceptional cases were then also excluded one by one. The only Voronoi cell that was left was the dodecahedral one. QED!

McLaughlin spent the fall of 1998 with Hales, putting the finishing touches to the paper, "A proof of the dodecahedral conjecture." On November 10, just two months after Hales announced his proof of Kepler's conjecture to the world, the proof of the dodecahedral conjecture was also in the bag. The following year, McLaughlin was awarded the AMS-MAA-SIAM[2] Morgan Prize for Outstanding Research in Mathematics by an Undergraduate Student, in recognition of this achievement. The prize carried a $1,000 cash payment and, more importantly, a certificate that stated that McLaughlin had received the most prestigious prize in mathematics that an undergraduate could win. In the citation the jury wrote that "the solution of this old, difficult conjecture constituted a singular achievement of such stature that this work alone was deserving of the highest recognition."

Next in line for Hales was the so-called honeycomb conjecture. In *The Six-Cornered Snowflake,* Kepler formulated not only the conjecture, which we discussed in the first thirteen chapters of the present book, but also one

---

[2] AMS—American Mathematical Society, MAA—Mathematical Association of America, SIAM—Society for Industrial and Applied Mathematics.

about floor tiling. Think of a tile layer by the name of Ernie, who has been commissioned to cover the floor of a large hotel lobby with tiles. The hotel owner will provide the tiles but Ernie has to provide the grout, which must fill the cracks where the tiles meet. The owner will be quite flexible about the shape of the tiles; all that matters to him is that they cover the whole lobby. In fact, tiles of different shapes would be okay too, and so would tiles that did not have straight edges. Ernie scratches his head. Business has been very slow lately, and he has to minimize expenses wherever possible. Furthermore, grout has become very expensive. What kind of tiles should he order? Without knowing it, Ernie is confronted with a two-thousand-year-old problem: What is the most efficient partitioning of the plane into equal areas?

In the local pub, after a couple of pints, Ernie tells his friends at the bar of his problem. As fate would have it, George, a bibliophile and antiquities buff, overhears the conversation. He immediately remembers a book on agriculture that he had heard about recently. It's a bit out of date, since it was written in 36 B.C. by the Roman scholar Marcus Terentius Varro. In this work Varro discussed the hexagonal form of the bee's honeycomb. Either the bees choose this shape in order to accommodate their six legs, he wrote, or there was some other reason. The other reason, he thought, was that this shape holds the largest amount of honey. "The geometers prove that this hexagon . . . encloses the greatest amount of space." George also tells Ernie about Kepler's *The Six-Cornered Snowflake*. In this booklet Kepler states that bees build their honeycombs in hexagonal patterns, because hexagonal walls require the least amount of wax. That's it for Ernie. Honey, wax, tiles, grout—it's all the same. He orders hexagonal tiles. Did he do the right thing?

Whether Varro's contemporaries actually proved that hexagons enclose the greatest amount of space is more than doubtful. In any case, no proof was ever found. Five centuries later, Pappus of Alexandria tried his hand at it. However, his "proof" was no more than a comparison of triangles, squares, and hexagons, the three regular shapes that tile the floor. Among these three shapes, the hexagon is the most efficient tile. But what about curved shapes, and what about joining different shapes, like a jigsaw puzzle? Then Charles Darwin, of all people, came up with his kind of proof. Since the production of wax costs energy, and bees, who have evolved over millions of generations, choose hexagons, then hexagons must represent the most efficient shape. But if mathematicians are weary of computers, they certainly won't accept evolutionary behavior as a valid proof. The question of the most efficient partitioning of space remained a puzzle for two millennia.

On Monday, August 10, 1998, Denis Weaire, a physicist from Trinity

Honeycomb tiling

College in Dublin, Ireland, was going fishing. He was just about to take his tackle out of the trunk of his car when his eye fell on a headline in a newspaper: "Kepler's Orange Stacking Problem Quashed." It was the day after Hales had announced to the world that he had solved Kepler's conjecture. Weaire, who had been working on a related problem for years, put his fishing tackle back. "All thoughts of angling were dismissed for a while," he would write later. In the ensuing e-mail correspondence he congratulated Tom on his achievement and then reminded him of the honeycomb conjecture: "Given its celebrated history, it seems worth a try."

That remark put Hales into high gear. What is the most efficient partitioning of the plane into cells of equal area? The question is reminiscent of Dido's problem (see chapter 3). First Hales studied the relevant literature and discovered, not totally to his surprise, that Laszlo Fejes-Tóth had been there before. In 1943 he had proved that among all straight-edged polygons, the hexagon comes out the winner.[3] But the prerequisite that polygons have straight edges was a very stringent assumption. After all, when Queen Dido wanted to encircle the greatest parcel of real estate with a strip of cow's skin, she founded the *round* city of Carthage. So why should circular edges be excluded? Fejes-Tóth didn't intend to exclude them at first, but soon found that "this conjecture resisted all attempts at proving it." So he made do with the weaker problem and predicted that the proof of the general problem would involve considerable difficulties.

Hales was intrigued. He started working on the problem in the winter of 1998. First he established an inequality that relates the area of a cell to its

---

[3] Actually, Fejes-Tóth proved the conjecture for convex cells. But convexity implies straight edges. (If bulges were allowed, then every cell with an out-bulge would necessitate another cell with an in-bulge, making it concave.)

perimeter. Then he showed that the area advantage of a cell whose sides bulge out is more than offset by the disadvantage of the neighboring cell whose sides bulge in. With that Hales had established that only polygons with straight edges could be optimal. Then he proved that the inequality that he established at the outset reaches its minimum when the cell is a hexagon. He completed the proof *without* any assumptions about the cells having straight borders. The following June, barely half a year later, he was done. Hales was surprised. The Kepler Conjecture had shaped his expectations of mathematical proofs: he had come to think that every age-old problem would require a monumental effort. He was unprepared for the light, twenty-page proof of the honeycomb conjecture he had found. "In contrast with the years of forced labor that gave the proof of the Kepler Conjecture, I felt as if I had won a lottery," he commented.

It is very rare that a mathematician succeeds in solving a centuries-old conjecture. To score a double whammy is quite unheard of. What's more, the proof of the latter conjecture was totally independent of the former, and was *not* a computer proof. Enough of telephone books, this time Hales had found an elegant proof.

It was only natural to move from partitioning the plane to partitioning space. But switching from two dimensions to three proved no simple task. The three-dimensional version of the honeycomb conjecture is called the Kelvin problem. It calls for the division of space into three-dimensional cells of equal volume, such that the total area of the walls is minimized. It is a very difficult problem. "Of course, I'm fascinated by the Kelvin problem too, but I don't think that it will be solved anytime soon," Hales wrote to Weaire.

Kelvin, born William Thomson in Belfast, Ireland, in 1824, was one of the most brilliant minds of the nineteenth century. His father was a professor of mathematics at Glasgow University, and he himself became a student at that institution at the tender age of 11. At the age of fifteen he won a gold medal for "An Essay on the Figure of the Earth," and his first scientific paper was published when he was sixteen. He continued his studies at Cambridge and in Paris. When the chair of natural philosophy—called physics today—at the University of Glasgow became vacant, his father mounted a carefully planned and energetic campaign to get his son nominated. He was successful and at the age of twenty-two, William Thomson became professor at the University of Glasgow. He remained there for his entire fifty-three-year career. Thomson's academic work covered thermodynamics, hydrodynamics, electricity, magnetism, and engineering. His contributions included six hundred scientific papers. In 1892 Queen Victoria raised him to the peerage. Sir William took the title of Baron Kelvin of Largs.

Toward the end of his life his insights became somewhat clouded. He opposed Darwin's theory of evolution, made incorrect speculations as to the age of the earth and the sun, opposed Rutherford's ideas of radioactivity, and maintained that there was nothing of importance left to discover in physics. "X-rays are a hoax," "heavier-than-air flying machines are impossible," and "radio has no future" are some of his pronouncements from that time. Nevertheless, his stature as a scientist was so established that when he died in 1907, he was buried next to Isaac Newton in Westminster Abbey.

Before we deal with the full Kelvin problem, let us inspect a partial version. Beehives should be efficient partitions of space. But at least one opening must be left in each cell, so that the bees can enter and leave their homes without disturbing neighbors. So what is the best design of cells, with one end left open? (The real Kelvin problem deals with cells that are closed on all sides.) The bees came up with a clever-looking plan: Walls were built along the hexagons, with four quadrangles added at the rear for privacy. The cells fit snugly back to back, and of course scientists, naturalists, and beekeepers immediately assumed that this design minimizes the amount of wax that is needed for the walls. So accustomed had one become to the idea that bees always do the right thing that it would have been "politically incorrect" to assume anything else. Bees just had to be superior mathematicians. By the way, with that, popular opinion had come full circle. While it had seemed incredible, at first, that a mere insect should be able to determine an optimal floor plan for its home, it now seemed unimaginable that bees did anything less than that in three dimensions. All that was missing was a proof. But in 1964 Fejes-Tóth thwarted all attempts to find one. In a paper entitled "What the bees know and what they do not know," he presented his own design for a bee cell. It saved a whopping 0.3 percent of wax. Whether

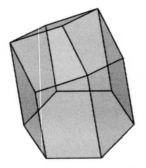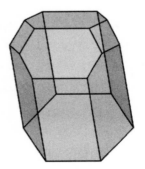

Bee's design (left) and Fejes-Tóth's design (right)

Fejes-Tóth's design actually is the absolutely best possible bee habitat has never been proven, but at least it served as a counterexample. Score one for *Homo sapiens.*

Back to the original problem. What is the best partitioning of space into cells of equal size, if there is no need for an entrance? Imagine a wall painter named Harry who must paint the rooms of a modern, Buckminster-Fuller-like house. All rooms have the same volume, and there are neither corridors nor lounges. One room just leads straight into the next. The builder is willing to construct the rooms in any shape or form, but Harry has to provide the paint for the walls, ceilings, and floors. Harry scratches his head. It wasn't only Ernie's business that had been slow lately, Harry too had to minimize expenses wherever possible. What room shapes should he ask the builder to construct so that he would require the least amount of paint?

In 1887 Lord Kelvin, then still Sir William Thomson, thought he had found a solution to just this problem. On September 29 of that year, at a quarter past seven o'clock in the morning, he woke up and jotted down an observation that had come to him during the night. (With that he placed himself firmly in the row of thinkers who work best while sleeping, or immediately after they wake up, like Gauss or Hales.) Kelvin had been wrestling with the new theories of light that James Clerk Maxwell (1831–1879) had proposed. According to this theory, which is correct, light is an electromagnetic phenomenon. But Kelvin was convinced otherwise. Until the end of his life he firmly believed that light emerges as the result of vibrations. But vibrations of what? It was the answer to that question that came to Kelvin during his dream. He conceived of a space-filling, invisible foam-like medium, which he called "ether." The immediate next question was: Of what shape could the foam's bubbles be? Kelvin thought that the wall area of the foam's bubbles must be minimized—since nearly everything in nature is either maximized or minimized—and gave the answer to the question in a paper entitled "On the division of space with minimum partitional area."

Kelvin built on the theories on soap bubbles of the Belgian physicist Joseph Antoine Ferdinand Plateau (1801–1883). Based on the realization that the energy of foam bubbles is proportional to its surface area, Plateau formulated three laws about their shapes and about how they can be connected to each other. His findings are all the more remarkable when one considers that this physicist was blind for the last forty years of his life. He had lost his eyesight after an experiment during which he stared straight at the sun for twenty-five seconds. From then on, the results of all experiments had to be described to him by family members, friends, and students. Plateau's laws led Kelvin to describe a slightly curved polyhedron con-

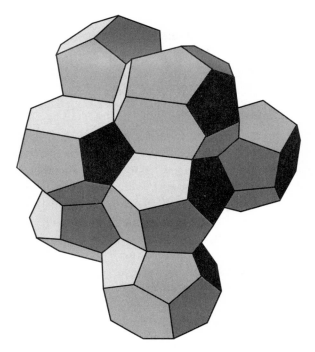

Kelvin's tetrakaidekahedron

sisting of six squares and eight hexagons. He called it the *tetrakaidekahedron* (TKD-hedron), literally the "four and ten seater." Copies of it fit together perfectly and the walls, ceilings, and floors require very little paint.

For more than one hundred years, Kelvin's TKD-hedron remained the best solution to this problem. His followers were ebullient; the master had solved yet another fiendishly difficult problem. Never mind that he had failed to provide a proof. (Never mind also that the ether that Kelvin proposed does not exist.) But soon a challenge arose. If nature is as clever as it pretends to be, and the TKD-hedron minimizes the surface of the walls, then surely TKD-hedra should appear somewhere in nature. Chemists, physicists, biologists, and other scientists started looking for this shape in their fields of study. No luck! The TKD-hedron failed to appear either naturally or in experiments. Gradually the euphoria gave way to disillusionment. Maybe TKD-hedra exist only in the virtual world. This left mankind with two possibilities: (1) either nature is not clever, or (2) the TKD-hedron is not the most economical partitioning of space. Which was it?

The Princeton mathematician Fred Almgren, an authority on soap bubbles, declared in 1982 that, "despite the claims of various authors to the contrary [Kelvin's conjecture] seems an open question." In 1993 Denis Weaire and his research student Robert Phelan got interested in the question. As scientists, they rejected option 1, of course. Nature cannot be anything but clever. Then they had an insight. Why should only a single cell be used to partition space? Maybe there exist two cells of equal volume but unequal shape that could be combined. They began hunting for the elusive wall-minimizing cells in nature. The cells must fit together without leaving gaps and should have less wall surface than Kelvin's TKD-hedra.

Weaire and Phelan let themselves be inspired by nature. In their quest to further human knowledge, they spent many hours at the Pavilion, one of the two places at Trinity College with a licensed bar available to staff and students. But while their colleagues peered deeply into their beer mugs—when they were not watching the cricket or rugby game from the Pavilion's veranda—Weaire and Phelan just skimmed the surface of their glasses. What they did, in fact, was inspect the foam on top. After about a month's work they hit paydirt. They identified bubbles that challenged Kelvin's TKD-hedron. While scrutinizing the foam, they also thought about analogous problems in the bonding structure of chemical compounds. Because nature is not only clever, but also quite repetitive. In Weaire's words it had been "one third intuition, one third analogy, and one third serendipity"—and one third beer, one is tempted to add.

But they still had a problem. In his first term as a graduate student, Phelan performed the computations that established the existence of a counterexample to Kelvin's conjecture.[4] So they knew more or less what the bubbles should look like. But they could not compute the exact surface of the walls. Neither could they determine their exact shapes. In their plight, Ken Brakke from Susquehanna University in Pennsylvania came to the rescue, like a knight in shining armor. Brakke had studied minimal surfaces, that is, soap bubbles, at Princeton University with Almgren's group. Subsequently he developed a computer program called Surface Evolver, which simulates and computes minimal surfaces. But instead of selling his program to breweries in search of the perfect foam, Brakke made the Evolver available to the scientific community. "My Surface Evolver is an interactive program for the modelling of liquid surfaces shaped by various forces and constraints. The program is available free of charge," he wrote on his web site.

The appropriate data was fed into their computers and—lo and behold—Brakke's program confirmed what Phelan and Weaire had suspected. The

---

[4] This guaranteed his eventual doctorate. Phelan now works for a Dutch telecommunications company.

Denis Weaire and Robert Phelan

Evolver produced two differently shaped cells: a pentagonal dodecahedron and a TKD-hedron. (The dodecahedron has pentagonal faces with sides of unequal length. The Weaire–Phelan version of the TKD-hedron is different from Kelvin's: it has twelve pentagonal and two hexagonal faces.) Mixed in the proportion of two of the former and six of the latter, they combine to form a structure that leaves no gaps in space. And now the punch line: The walls of the structure have 0.3% less surface area than Kelvin's structure. This may seem like a negligible improvement, but consider that the development of the world records in the 100-meter dash or the long jump are characterized by even smaller increments. And then, Fejes-Tóth's improvement over the bee cells was just as minute.

Informed of the discovery, Brakke immediately replied with an e-mail confirming their success: "As soon as I saw the picture on the screen, I was sure you had it. . . . Congratulations." Almgren exclaimed that this was "a glorious day for surface minimization theory." With Weaire and Phelan's discovery, Kelvin's century-old conjecture had been proven incorrect. Were he alive today, Kelvin could at least take comfort in the thought that bees had missed the optimum by about the same margin.

There is no guarantee that the Weaire–Phelan partition is optimal. Pending such proof, scientists are still trying to come up with bubbles that are even more economical partitions of space. So far, nobody has had any luck. In the meantime, Weaire was made a Fellow of the Royal Society of London with a citation that includes the discovery of the space partition.

Another problem that was recently solved concerns the so-called double-bubble: What shape do two bubbles with given volumes have after they dock? We already mentioned in chapter 1 that the problem goes back to Archimedes, who conjectured that a round sphere is the most economical way to enclose a given volume, and that Hermann Amandus Schwarz proved the conjecture in 1884.

It then took another century until a team of mathematicians managed to tackle the double-bubble. We know that soap bubbles minimize surface, so the task consists of finding the shape of two combined bubbles that minimizes the total surface. (Astonishingly, a group of undergraduates at Williams College solved the two-dimensional version of the problem in 1990.) In three dimensions, a computer proof had already been proposed as a partial solution. But it was unsatisfactory. Then, in March 2000, four mathematicians—Michael Hutchings, Frank Morgan, Manuel Ritoré, and Antonio Ros—presented a traditional, elegant proof for the general problem. They showed that two bubbles must dock in the obvious way, side by side. Other wild configurations, for instance one bubble wrapped around the other, were shown to be unstable. In real soap-bubble life, they would just go pop! Then it was the undergraduates' turn again. A new group of youngsters spent a summer at Williams College under the direction of Frank Morgan, and extended the proof to four-dimensional bubbles.

These are some of the success stories, but there remain numerous unsolved problems. One of them goes by the name "sausage conjecture." It is related to Kepler's conjecture and was proposed by that Hungarian grandmaster of discrete geometry, Laszlo Fejes-Tóth. To this day it remains unsolved. To illustrate, let us observe Fumiko, a shop assistant in Tokyo who

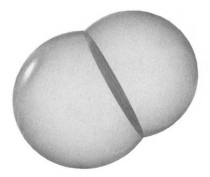

Double-bubble

boxes up gifts for the customers in the sports department. The question she faces is, what is the best way to pack balls? Which boxing method wastes the least amount of space? For example, is it more efficient to pack four tennis balls—or Ping-Pong balls, or golf balls—into a square box, or an elongated box or a pyramidal box? How about six balls, or three, or twenty-five? We will call the long boxes sausage-boxes and all others cluster-boxes.

The results are quite surprising. For up to fifty-five balls the sausage seems to represent the best box. It may be a little awkward to load such a thing into the trunk of the car, but as far as wasted space is concerned it's the best. If you want to pack additional balls something strange happens, however. For fifty-six balls or more, clusters are better than sausages. Fejes-Tóth was so shocked and confused when he found out that he called the switch from sausages to clusters the "sausage catastrophe." Nobody is quite sure at exactly what number of balls optimality changes from sausage to cluster. It is believed that the switch occurs somewhere between fifty and fifty-six balls. But at exactly what number this happens is still an open question.

If you become bored with regular tennis you may prefer to play the game in four dimensions. As you might expect, the problem does not become any easier. As long as you play four-dimensional tennis with less than about seventy-five thousand balls, it is better to store them in sausage-boxes. For more than 375,769, cluster-boxes are best. Somewhere in between the switch takes place. Where exactly? Nobody knows.

Surprisingly, in still higher dimensions the situation calms down again. Fejes-Tóth claimed that in sufficiently high dimensions, sausages are always best, no matter how many balls you want to store. This contention was proved rigorously for all dimensions greater than forty-two. (Quite a tennis

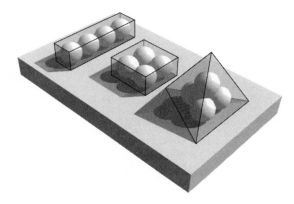

Sausage-box or cluster-boxes

game that is: You don't just have to look ahead at your opponent but must keep your eyes peeled in forty-two directions simultaneously.) But Fejes-Tóth conjectured that in any dimension higher than five, sausages are always best. As of this writing, the claim is still awaiting a proof. By the way, can you guess what happens in two dimensions? It's real simple: clusters are always superior to sausages.[5] And in one dimension, sausages are the only option.

If the surface area of the box's walls are of primary concern and not the volume, then there is another conjecture in store. The "spherical conjecture" states that the optimum, wall-minimizing shape of the box for tennis balls is roughly spherical if the dimension is large. Whether this is true, and for which dimensions the conjecture should hold, is anybody's guess at this stage.

And then there is Tammes's problem. P. M. L. Tammes was a Dutch botanist who investigated why the pores on pollen grain are distributed in a regular fashion over the grain's surface. He surmised that they are so arranged to be as far away from each other as possible. This is equivalent to decorating a sphere with curved disks that must not overlap. What is the maximum radius the disks can have? In his 1930 paper "On the origin of number and arrangement of the places of exit on the surface of pollen grains," Tammes discussed this question. He also proved an interesting fact: It does not matter whether there are five pores or six, the maximum radius is the same in both cases. Of course, the botanist's proof wasn't rigorous enough for mathematicians and was corrected sixty-six years later. But the general problem, for any number of orifices, remains with us even today. There is an enormous literature on this problem but exact solutions are known only for twelve points or less, and for twenty-four points. For any other number of points only bounds on the maximum radii have been constructed.[6]

But what are all these theorems and proofs good for? What about applications? In mathematical circles such questions are considered very gauche, if not downright rude. It is like asking a mountain climber why he climbs mountains. George Leigh Mallory's answer to that question may sum it up also for mathematicians: he climbed a mountain "because it's there." Maybe he should have taken the question more seriously before giving a snappy

---

[5] Recall the coins on the tabletop.
[6] Tammes's problem is related to the kissing problem of chapter 5. The question there was whether thirteen points could be placed on a sphere, such that they would be separated from each other by a distance of at least $2\pi/6$.

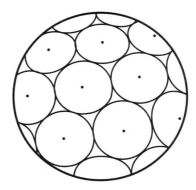

Tammes's problem

answer. His frozen body was discovered on Mount Everest only seventy-five years after he disappeared on a 1924 expedition.[7]

Mathematics certainly is no extreme sport, but mathematicians also solve problems just because they're there. They pursue the work for the subject's intrinsic beauty. The joy of mathematics consists in doing it—gratification is mathematics itself. But occasionally one must explain what one does for a living to one's mother, or justify one's work to a funding organization, or impress one's boy- or girlfriend. In these cases it does not hurt to have some real-life examples handy. Take John von Neumann. He could always mention game theory, atom bombs, or electronic computers, depending on who was listening. But what can packing experts show for their labor? How to stack oranges? You won't be able to dazzle a date with that. And you would be hard-pressed to find a funding agency willing to underwrite research on how to improve the stacking of melons. In the next—and final—chapter I will describe some areas where the theory of sphere packing can be applied.

---

[7] The question of whether Mallory had reached the summit before he fell to his death, twenty-nine years before Sir Edmund Hillary reached the top, or on his way up, has never been settled.

CHAPTER 15

This Is Not an Epilogue

L et us return to Thomas Harriot and Johannes Kepler at the end of the sixteenth century to describe some applications of the theory of densest packings. Harriot, moving beyond Raleigh's cannonballs, wondered how atoms are arranged around each other. This is quite remarkable, since atoms were no more than a figment of imagination at that time. Kepler, for example, did not even believe that these little "balls" existed. But Harriot was right on target. The packing of spheres in three-dimensional space serves as a correct model for the understanding of how matter is built.

Packings are especially helpful to the understanding of the structure of crystals.[1] Most metal atoms, for example, are arranged either in an FCC or an HCP structure. As Harriot speculated, that is one reason why metals are heavier than other materials: they are more densely packed.[2] Besides the FCC and the HCP, other crystalline structures exist: the simple cubic packing (SCP), where eight atoms sit at the corners of a cube, and the body centered cubic packing (BCC), which is the same as the SCP with an additional atom sitting in the middle of the cube.[3] Obviously, these two structures are less dense than FCC and HCP. After all, the whole purpose of Tom Hales's exercise was to show that the density of Kepler's arrangements (74.05 percent) is the highest possible.

The SCP packing is a very inefficient packing. Atoms fill only 52 percent of the volume (see the appendix to chapter 1) and only one chemical element exists whose atoms conform to this arrangement: the radioactive

---

[1] In general parlance, crystals are associated with quartz crystals or crystal glass. Not so in chemistry. There, crystals describe the state of a chemical element whose atoms are arranged in a periodic lattice.

[2] The two other explanations for an element's mass are the number of protons in each atom and the atom's size.

[3] There are more crystal structures, but we won't deal with them here.

polonium, discovered by Pierre and Marie Curie in 1898. The BCC, at 68 percent, is somewhat denser. Chromium, sodium, and iron atoms, for example, conform to the BCC packing. Cadmium, cobalt, and zinc are instances of the HCP packings. Finally, the queen of packings, the FCC, is adopted by—what else—silver, gold, and platinum.

The sphere packing model also indicates why some elements are more flexible than others. When you deform a material you actually push the planes of atoms over each other. Atoms in the BCC structure fit into the interstices formed by four atoms of the lower layer. In the FCC structure an atom rests in the interstice formed by only three atoms of the lower layer. Hence BCC atoms fit more deeply into the cracks and it is more difficult to budge them out of their nests. That is why BCC atoms, like chromium, are usually firmer than FCC structures, like gold.

The sphere packing model also enlightens us about the advantages of impurities in some materials. Defects in the lattice operate like cogs; they prevent slipping. And that is why the "pure" gold earrings you just bought at the jewelers are at best only 75 percent pure and more likely only 58 per-cent (18 and 14 carats, respectively). The remainder consist of silver, cop-per, or some other metal. Twenty-four-carat gold would be too malleable.

Predictably enough, packing theory can also be used to solve problems of how to pack items into containers. Applications include the bin packing problem (put as many objects as possible into the least number of contain-ers), the dual bin packing problem (fill as many bins as possible with the fewest objects), the knapsack problem (given items of different values and volumes, find the most valuable pieces that fit into a knapsack of fixed vol-ume), the cutting stock problem (find the arrangement of shapes on a sur-face that minimizes waste), or the strip packing problem (cut as many strips of specified lengths as possible from a number of fixed-length strips). In each of these examples the objects must be packed as densely as possible, thus minimizing wasted space.

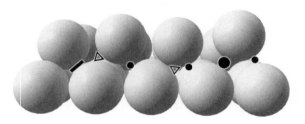

FCC structure with deforming material

These examples come from an area of mathematics called operations research. In practice they appear, for instance, in the stacking of containers onto ships and the loading of pallets onto airplanes and trucks. The problems are very difficult. In the old days freight forwarders just loaded goods helter-skelter onto the platform. When additional items would no longer fit, the ships, trucks, and planes took off, much to the chagrin of the owners, whose vessels and vehicles were usually still partially empty. The freight forwarders didn't care, of course, since they paid by weight. But then the powers-to-be had a good idea. They decided to standardize containers ($8' \times 8.5' \times 20'$ or $8' \times 8.5' \times 40'$) and pallets ($48'' \times 40''$ in the United States, and 120 cm $\times$ 80 cm in Europe; height is usually not specified). Once containers and pallets were built to these sizes, the boxes fit neatly next to, and on top of, each other. Ships were built to specifications and containers could be loaded into the hold and onto the deck without wasting any space. Thus we have again come full circle. Remember Sir Walter Raleigh? He started the whole saga by wondering about the best way to load cannonballs onto a ship.

With the standardization of the containers the ship owners had rid themselves of a serious problem. But the problem of wasted space hadn't disappeared into thin air, of course. It had simply been shifted downwards to the freight forwarders. They were the ones who now had to worry about how to pack their wares as densely as possible into their containers. Ship owners had the last laugh.

But it is not always possible to standardize. The job of a tailor, cutting cloth for shirts and pants, would be much easier if the garments were made of standardized rectangular pieces of cloth.[4] But variation is the name of the game in many industrial applications, like sheet metal factories, paper mills, and garment sweatshops. Plasticine can always be resquashed and reused. Gold and silver can be smelted and reused. Not so the expensive lacework that must be trimmed from the brocade. Too much waste cuts into the profits. Leather cutting presents even more difficult problems, since imperfections in animal hides must be avoided.

Another illustration of a packing problem is the placing of fishing rods, golf clubs, tennis racquets, and sausage- and cluster-boxes with golf and tennis balls into the trunk of your car. The decisions you have to make when buying a bottle of cola at a vending machine, knowing that you need change for the highway toll and another quarter to tip the guy who washes the car windows at the red light, also belong to the general realm of packing problems. So do newspaper layouts and the ancient Chinese puzzle Tangram, or

---

[4] Even then the problem is by no means trivial.

its three-dimensional version, the Soma cube. There is no general method to solve these problems. Algorithms have been developed that are designed to provide optimal solutions for many specific problems. In the simplest cases the Simplex algorithm may give a solution, but usually much more sophisticated programs are required.

Finally, there are some farfetched applications of packing problems. Take telecommunications. Let's say a signal consists of a string of ten digits between zero and nine, like 3849001823 or 8640923902. Hence, each signal is represented by a point in a ten-dimensional cube with edge-length 10. Altogether there is room in this hypercube for 10,000,000,000 different signals. But add a constraint. Since similar strings could be confused because of noise in the transmission lines, we won't allow signals that are "close" to each other, like 1234567890 and 1234567891. Only strings that differ in each position by at least two units are allowed. Then, if a slightly distorted, and therefore illegal, message arrives at the receiving end, it can be rectified by simply assigning it to the closest legal string. The meaning of this constraint is that every signal that lies within a sphere of radius one around the original signal is not legitimate.[5] This just begs the question: Given this constraint, how many different signals can be represented with strings of length 10? The latter is tantamount to the question: What is the densest packing of spheres with radius 1, in a ten-dimensional hypercube? This application of the theory of sphere packing goes under the name "error correcting codes." It is suspected, but not known, that the best density of spheres in ten-dimensional space is just under 10 percent. Hence 400,000,000 signals can be represented, which is sufficient for all words in all languages of the globe.[6] But if the radius of the spheres is increased to four units, in order to make errors even more unlikely, a ten-dimensional signal allows only 374 words or so.[7] So a higher-dimensional hypercube for coding may be needed. How high the dimension must be depends crucially on how densely spheres can be packed. Hales provided the answer for three dimensions. For dimensions higher than three the answer is unknown.

Let's describe another farfetched example: running a business. One of the first things MBA students learn in business school is that profits must be maximized. The second thing they learn is that to do so, markets should be segmented. Big spenders should be charged high prices, but cheapies

---

[5] Actually, this defines a cube, but we won't go into details.
[6] The volume of a ten-dimensional sphere of radius 1 is about 2.55. Ten percent of ten billion divided by 2.55 is about 400 million.
[7] A ten-dimensional sphere of radius 3 has a volume of 2,674,041. Ten percent of ten billion divided by 2,674,041 is about 374.

should not be disregarded. An illustration of this maxim is available in most commercial airlines. Usually there's a first class in front, followed by business class, and then there's economy in the back. The "product" is the same (transporting a person from, say, London to New York), but those who can afford it should pay a higher price than those who cannot. Accordingly, advertising campaigns will emphasize the product's different virtues to different potential clients. A marketing-savvy garment supplier will stress to its young buyers how cool it is to wear a necktie, while it will put the emphasis on a necktie's elegance to older buyers. A tennis-shoe company may sell the same product to men and to women, but package them in different colors. Sometimes a market must be segmented according to more than one criterion: age, sex, income, education, and so on. For each niche a different advertising strategy may be designed: one for highly educated women between the ages of forty-five and fifty-five with incomes over $100,000 per year, another for young men without high-school diplomas earning less than $25,000, and so on. The vice president of marketing wants to reach as many potential customers as possible, spending as little as possible, without ignoring anyone, and without any overlap. And here is the link to sphere packings: The allocation of advertising budgets in segmented markets is equivalent to the packing of spheres in high-dimensional spaces.

Believe it or not, the theory of sphere packing can also be applied to political science. Political parties must decide how they should position themselves in the space of potential voters, minimizing both residual space and overlap. For example, a lattice packing can reflect a copying mechanism in the positioning of political parties ("copy the neighboring parties, but differentiate your platform at least in one dimension"). As the dimensions increase, pockets open up between the constituencies that were traditionally covered by the large consensus parties. That is where small, special-interest politicians may find a foothold.

With this our account comes to an end. But the story does not, because mathematics is never completed. The speed with which questions, problems, and hypotheses are solved is surpassed by the pace at which new ones arise. In the appendix, a few dozen conjectures are listed that have been proposed over the years. Some of them have been solved, more may be solved by the time you read this, but most will still be looking for a solution for a long time to come. Therefore, let us not consider this chapter the epilogue to one story, but rather a prologue to new endeavors.

Volume of *n*-Dimensional Spheres

Dimension	Sphere	Volume	Volume for $R = 1$ cm
1	Line joining the two endpoints	$2R$	2 cm
2	Circle, and all points inside it	πR^2	3.14 cm²
3	Solid ball	$\frac{1}{3}\pi R^3$	4.19 cm³
4	4-dimensional solid ball	$\frac{1}{2}\pi^2 R^4$	4.93 cm⁴
⋮	⋮	⋮	⋮

Density of Coins in the Plane

The surface of a triangle is given by multiplying its base by its height, and dividing the result by two:

$$\text{Surface of triangle} = \frac{Base \cdot Height}{2}$$

The surface of a circle is the number π multiplied by the square of the radius:

$$\text{Surface of circle} = \pi r^2$$

(We usually will assume in this book that the spheres' radii, r, are equal to 1.) It follows that the length of each of the triangle's edges is 2. Since a surface can be tiled with identical triangles, it suffices to inspect just one such triangle.

Pythagoras's Theorem tells us that the height of an equilateral triangle, with edge-length 2 is $\sqrt{3}$. Hence our triangle's surface is

$$\frac{Base \cdot Height}{2} = \frac{2\sqrt{3}}{2} = \sqrt{3} \approx 1.732$$

The density is defined as the ratio of (1) the surface of the triangle that is covered by circles to (2) the total surface of the triangle. Now *one-sixth* of each of *three* circles is contained in this triangle. Hence the parts of the spheres' surfaces that are contained within the triangle are

$$\frac{1}{6} \, 3(\pi r^2) \approx 1.571$$

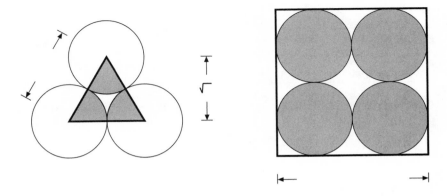

Recall that the radius $r = 1$. Hence the proportion of those parts of the spheres' surfaces that are contained in the triangle, to the surface of the triangle itself, is

$$Density \approx \frac{1.571}{1.732} \approx 0.907$$

In other words, nearly 90.7 percent of the triangle is covered by the spheres. And since the plane can be regularly tiled with identical triangles, this is also the density of the plane.

Density of the Regular Square Packing

Take a square with a side length of 4. Its surface is 16. Four whole circles can be fitted into such a square, and their combined surface is

$$4(\pi r^2) \approx 12.566$$

Hence the density of the packing within that square can be computed as

$$Density \approx \frac{12.566}{16.0} \approx 0.785$$

or about 78.5 percent. Since an infinite surface can be tiled with identical squares, the density of the square is identical to density of surface extended to infinity.

The Melon Rind

The following table gives surfaces and volumes of round melons (with radius $r = 1$) and of cube-shaped melons (with edge-length s):

	Surface	Volume
Round	$4(\pi r^2) \approx 12.566$	$\frac{4}{3}(\pi r^3) = 4.189$
Cube-shaped	$6s^2$	s^3

What edge length must a cube-shaped melon have in order to contain the same volume as the round melon with radius $r = 1$? Let us set the volumes of both kinds of melons equal:

$$4.189 = s^3$$

We solve for s, to get

$$s = \sqrt[3]{4.189} \approx 1.612$$

Hence a cube-shaped melon with edge-length 1.612 has the same volume and weight as a round melon with radius 1. According to the above table the skin surface of a cube-shaped melon with edge-length $s = 1.61$ is

$$6s^2 \approx 6(1.612)^2 \approx 6(2.599) \approx 15.591$$

while the rind of the round melon is only 12.566. Hence round melons require nearly 20 percent less rind than the cube-shaped ones.

Density of Melon Heaps

In the first part of the appendix we looked at the regular square packing in two dimensions. Let us now analyze the same packing in three dimensions. Consider cubes with edge-lengths 2. (Since the whole space can be completely filled with identical cubes, it suffices to inspect just one such cube.) Each of these cubes has a volume of $2^3 = 8.0$, and exactly one sphere fits into each such cube. Since the volume of a sphere is

$$\frac{4}{3}(\pi r^3) = 4.189$$

the density of this packing is $^{4.189}\!/_{8.0} \approx 0.52$, that is, 52 percent.

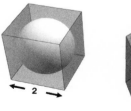

But we can do better than that. We already know that in two dimensions the hexagonal arrangement allows more coins to be packed than the square arrangement. Let us see what happens if we replace the coins by melons and add layers. We extend the hexagons into three dimensions by making them into cylinders of height two. Now we can fit one sphere into each of these hexagonal cylinders. Let us compute the density of this arrangement. The surface of a hexagon is

$$Surface(Hexagon) = 2\sqrt{3}r^2$$

where r is the radius of the inscribed circle, that is, the distance from the center of the hexagon to its edge. The volume of a hexagonal cylinder with height 2 is (recall $r = 1$):

$$Volume(Hexagon) = 2(2\sqrt{3}r^2) \approx 6.928$$

The volume of the melon is 4.189 (see above), hence the density of this packing is $^{4.189}\!/_{6.928} \approx 0.605$, or 60.5 percent. Since all additional layers are identical to the first one, this is also the density of the hexagonal packing in three dimensions. Obviously 60.5 percent is better than the 52 percent of the square packing. But it is not as good as the density of the FCC and the HCP (74.05 percent), where the melons of each additional layer are placed in the dimples that are formed by the melons of the previous layer. This we will see in the appendix to chapter 2.

In this appendix we show that the square or hexagonal packing of spheres in three dimensions fills 74.05 percent of the space. In order to do so we first partition space into equal cells, such that each sphere lies inside one such Voronoi cell. (Voronoi cells are discussed in detail in chapter 9.) What is the shape of such a cell? If the arrangement of the spheres were a simple square packing, then the Voronoi cells would be cubes stacked on top of, and next to each other. But since the spheres in Kepler's arrangement lie in a more complicated arrangement—the balls lie in the dimples that are formed by the balls below, and each ball is touched by twelve others—the cells also have a more intricate shape. The Voronoi cells that surround the balls are shaped as so-called *rhombic dodecahedra*.

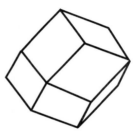

Rhombic dodecahedron

In order to compute the density of one cell, all one has to do now is to compute the volumes of the ball and of the rhombic dodecahedron that surrounds it. Since space can be filled to infinity with rhombic dodecahedra without leaving any gaps, the density computed for a single ball in its Voronoi cell also holds for the packing in infinite space. This sounds simple enough but, unfortunately, there is a slight problem: Computing the volume of the Voronoi cell for Kepler's packing is no easy procedure. I will just state here that the rhombic dodecahedron (for balls of radius $r = 1$) has a volume of $4\sqrt{2}$, which equals about 5.6568. . . .

On the other hand, the volume of a ball with a radius of 1 is $\frac{4}{3}\pi$, or 4.18879. . . . We now compute the density of the packing by dividing the volume of the ball by the volume of the rhombic dodecahedron, $\frac{4.1888}{5.6568}$, which equals 0.74048 . . . , or about 74.05 percent.

The Quadratic Form and the Diagonals of the Fundamental Cell

What is the length of the short diagonal (d_1) through the fundamental cell? In the picture we add the height (h) of the fundamental cell, and divide vector c into two segments, p and q. By Pythagoras's Theorem we know that the square of d_1 is equal to the sum of the squares of h and q:

$$d_1^2 = h^2 + q^2$$

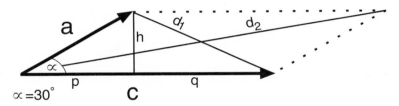

Quadratic form and the diagonals

We can use Pythagoras's Theorem once more to determine h^2:

$$h^2 = a^2 - p^2$$

From the picture we see that $q = c - p$. Hence we have

$$d_1^2 = a^2 - p^2 + (c - p)^2$$
$$= a^2 - p^2 + c^2 - 2cp + p^2$$
$$= a^2 - 2cp + c^2$$

Let's denote the angle between the two vectors by α. Unfortunately, angles have the annoying habit of being measured in degrees, and to avoid complications we introduce a new variable, denoted by b, that is measured in real numbers:

$$b = a \cdot c \cdot \cos \alpha$$

We can rewrite this as $\cos \alpha = b/ac$. But $\cos \alpha$ also equals p/a, so $p = a \cos \alpha$. Entering b/ac for $\cos \alpha$ we get $p = b/c$. Entering this in the above equation, we get

$$d_1^2 = a^2 - 2b + c^2$$

So the square root of the quadratic form is just the length of the short diagonal. And this is just what we set out to prove. In a similar manner we can show that the length of the second diagonal, d_2, is equal to the quadratic form with the sign of the middle term changed to a "plus":

$$d_2^2 = a^2 + 2b + c^2$$

The Discriminant and the Surface of the Fundamental Cell

The surface of the fundamental cell, denoted by F, is equal to its base times its height, that is, $F = c \cdot h$. Hence we have

$$F^2 = h^2 c^2$$

We know from basic trigonometry that

$$\sin \alpha = \frac{h}{a} \qquad \text{i.e., } h = a \sin \alpha$$

Pythagoras's Theorem tells us that

$$h^2 + p^2 = a^2$$

or

$$h^2 = a^2 - p^2$$

From $\cos \alpha = p/a$, we have $p = a \cos \alpha$, and therefore we get

$$h^2 = a^2 (1 - \cos^2 \alpha)$$

From the definition of b it follows that $\cos \alpha$ equals b/ac. Putting all of this together, we obtain

$$F^2 = a^2 c^2 \left(1 - \frac{b^2}{a^2 c^2} \right)$$

or, after rearranging,

$$F^2 = a^2 c^2 - b^2$$

The right-hand side of the last equation is the discriminant of the quadratic form, and we have therefore shown that its square root is just the surface of the fundamental cell.

Examples

To illustrate, we return to Manhattan's grid layout. The angle between avenues and streets is 90°, and since the cosine of a right angle is zero, b is also zero. Hence the quadratic form is $300^2 + 0 + 100^2$. The square root of this is 316 meters, and this is the length of the diagonal through a block. The discriminant associated with this quadratic form is $300^2 \cdot 100^2 - 0$, and the surface of a block is the square root of this number, that is, 30,000 square meters. Quadratic forms and their discriminants can be used for any lattice, after the appropriate values for a, b, and c have been inserted.[1]

Lagrange also provided formulas that can be applied for areas, which cover more than just one street and one avenue. Let us look at a *superblock* stretching for X avenues (each having length a) and Y streets (of lengths c). First, the diagonal distances through this superblock are equal to the square root of the quadratic forms, $a^2X^2 + 2bXY + c^2Y^2$ and $a^2X^2 - 2bXY + c^2Y^2$, and the surface is the square root of the discriminant $a^2c^2X^2Y^2 - b^2X^2Y^2$. Once a, b, and c are known, all one has to do to get the lengths of the two diagonals and the surface is enter X and Y into the formulas. The length of the diagonals through just one block can easily be computed by setting X and Y equal to 1. The length of the diagonals through four blocks (for example, from 3rd Avenue and 50th Street to 5th Avenue and 52nd Street) can be computed by setting X and Y equal to 2 in the formula. We leave the verification—that the diagonal of the superblock is 632 meters and its surface is 120,000 square meters—as an exercise for the reader.

Suppose we have circles with a 50-meter radius that we want to place around Manhattan. Both the sides of the blocks (100 and 300 meters) and the diagonals (316 meters) are sufficiently long to easily place circles at the four corners without any of them overlapping. Hence the Manhattan lattice is a contender for the title of closest packing. The discriminant, $a^2c^2 - b^2$ equals 900,000,000, and the square root of this, which is just the surface of the block, is 30,000 square meters.

Let us check how well the Manhattan lattice performs. According to Lagrange, the discriminant can never be smaller than $\sqrt{(\frac{3}{4}a^4)}$, which equals $0.866a^2$ or 8,660 square meters. So we didn't expect any surface smaller than that. But the Manhattan block does not even get close. Maybe we can construct a better grid by changing the angle between the two vectors. Let us compute the surface of a *deformed* Manhattan lattice by reducing the angle between the two vectors to about 70°. This corresponds to a value for b of 10,000. The first thing to check is whether the diagonals are longer

---

[1] When α is 90° we could simply use Pythagoras's Theorem to compute the diagonal of the cell. The significance of quadratic forms is that the diagonal of *any* cell, regardless of its angle, can be computed. (By the way, the surface of a *superblock* is xy times the square root of the discriminant.)

than 100 meters. Only then can circles of 50-meter radius be fit into the corners without overlap, and only then is the deformed lattice a legitimate contender. The square roots of the quadratic forms give the lengths of the diagonals of a block:

$$\sqrt{(10{,}000 + 2(10{,}000) + 90{,}000)} = \sqrt{120{,}000} \approx 346 \text{ meters}$$

$$\sqrt{(10{,}000 - 2(10{,}000) + 90{,}000)} = \sqrt{80{,}000} \approx 283 \text{ meters}$$

Hence, the diagonals are sufficiently long to fit circles comfortably into the four corners of the block of the deformed Manhattan lattice. Now what's the surface of such a block? The discriminant, $a^2c^2 - b^2$, equals

$$300^2 100^2 - 10{,}000^2 = 900{,}000{,}000 - 100{,}000{,}000 = 800{,}000{,}000$$

The square root of this number, and hence the surface of the block, is 28,284 square meters. The deformed Manhattan block has become somewhat smaller than 30,000 m^2, but it is still a long, long way off from 8,660 m^2. Let's try a deformed Manhattan lattice but shorten the avenues. By how much can the avenue vector be cut? Obviously the vector can not be shorter than 100 meters, because there must be enough room for two circles next to each other. So the shortest possible avenue has the same length as the street, that is, $a = c$. Add to that (without proof) an angle of 60° and we have the hexagonal arrangement.

Out of curiosity, let us check the density of such a packing. The surface of a circle with a radius of 50 meters is $\pi \cdot 50^2 \approx 7{,}854$ m^2. Divide this by the surface of the smallest cell imaginable, which is 8,660 m^2, and one gets $7854/8660 = 90.69\%$. Wow, that is just the density of the closest packing. (See the appendix to chapter 1.) Well, maybe that isn't all that surprising since we already know that among all possible lattices, the hexagonal one allows the densest arrangement of circles in a plane.

Circle around an Equilateral Triangle

Consider an equilateral triangle of side length D. We want to draw a circle that runs through the three corners of the triangle. Obviously, the centers of the circle and of the triangle coincide. What is the circle's radius?

The height of the triangle, H, can be computed with the help of Pythagoras's Theorem, as

$$H = \sqrt{D^2 - \left(\frac{D}{2}\right)^2} = \frac{\sqrt{3}}{2} D$$

The triangle's height is composed of two components, the radius that we are seeking, R, and another segment that we shall denote by S:

$$H = R + S$$

Hence

$$R = \frac{\sqrt{3}}{2} D - S \qquad (1)$$

We can use Pythagoras's Theorem once more to get

$$R^2 = S^2 + \left(\frac{D}{2}\right)^2 \qquad (2)$$

Expressing equation (1) as

$$S^2 = \left(\frac{\sqrt{3}}{2} D\right)^2 - \sqrt{3}DR + R^2$$

and substituting into equation (2), we get

$$R^2 = \left(\frac{\sqrt{3}}{2} D\right)^2 - \sqrt{3}DR + R^2 + \left(\frac{D}{2}\right)^2$$

In this equation the terms R^2 cancel, and we obtain

$$D^2 - \sqrt{3}DR = 0$$

from which follows

$$R = \frac{D}{\sqrt{3}}$$

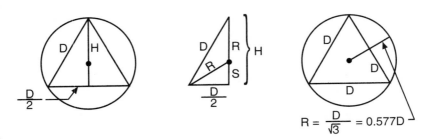

Circle around an equilateral triangle

Hence the radius of the circle around an equilateral triangle with side length D is $D/\sqrt{3}$, which equals $0.577D$.

Side Lengths of Octagons and Heptagons

We consider a circle of radius $2D/\sqrt{3}$. First we inscribe an octagon, then we will do the same exercise for a heptagon. An octagon can be partitioned into eight triangles. Obviously the angle of each triangle in the center is $\frac{360°}{8} = 45°$. Therefore one half of the octagon's side length, C, is

$$\frac{C}{2} = \sin\left(\frac{45°}{2}\right)\frac{2D}{\sqrt{3}} = 0.383\,\frac{2D}{1.732} = 0.442D$$

The octagon's side length is twice this length, that is, $C = 0.884D$. This is too short! Two adjacent points do not have the required distance of D from each other.[1]

Let us move on to the heptagon. The angle of each of the seven triangle in the center is $\frac{360°}{7} = 51.43°$. Therefore one half of the heptagon's side length, G, is

$$\frac{G}{2} = \sin\left(\frac{51.43°}{2}\right)\frac{2G}{\sqrt{3}} = 0.434\,\frac{2G}{1.732} = 0.501D$$

The heptagon's side length is twice this length, that is, $G = 1.002D$.

---

[1] We must also deal with the case of star-shaped octagons, where the corners lie on the outer and inner edge of the ring. Calculations show that this reduces the distance between any two adjacent corners even more.

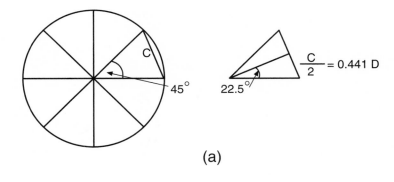

(a)

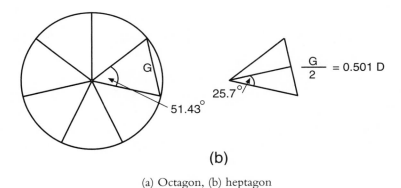

(b)

(a) Octagon, (b) heptagon

Surface of a Hexagon

Consider an equilateral triangle of side D. As was shown in the first part of this appendix, the radius of the circle that cuts through the three corners of the triangle has a radius of $D/\sqrt{3}$. This is the side length of the hexagon that we must now consider.

What is the surface of a hexagon with side length K? The hexagon can be partitioned into six triangles, and the surface of each triangle can be computed as follows. First compute the height of the triangle, H:

$$H = \sqrt{K^2 - \left(\frac{K}{2}\right)^2} = \frac{\sqrt{3}}{2} K$$

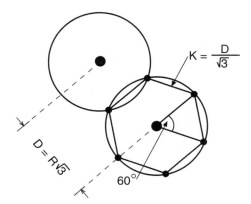

$$K = \frac{D}{\sqrt{3}}$$

$$D = R\sqrt{3}$$

$$60°$$

Surface of a hexagon

Since the surface of a triangle is equal to half the base times the height, we get

$$Surface_{Triangle} = \frac{1}{2}\left(\frac{K}{2}\right)H = \frac{3\sqrt{3}}{4}K^2$$

Hence the hexagon's *six* triangles cover an area of

$$Surface_{Hexagon} = 6\frac{\sqrt{3}}{4}K^2 = \frac{3/\sqrt{3}}{2}K^2$$

We are considering hexagons with a side length $K = D/\sqrt{3}$. Hence, the hexagons in Fejes-Tóth's proof cover an area of

$$Surface_{Fejes\text{-}Tóth} = \frac{3\sqrt{3}}{2}\left(\frac{D^2}{3}\right) \approx 0.866\,D^2$$

Superball and the Shadows of Kissing Balls

From high school we know (and if we don't, we can look it up in any collection of mathematical formulas) that the surface of a sphere's cap is given by

$$C = 2\pi r(r - h)$$

where $r = 3$ and $r - h$ is the height of the cap. How large is h? By Pythagoras's Theorem we have

$$h = \sqrt{r^2 - x^2}$$

where x is the diameter of the cap, which needs to be computed. From chapter 4 we know that six circles can touch the central circle in two dimensions, hence the angle that encompasses each circle spans 60°. The sine of half that angle allows the computation of x:

$$\sin 30° = \frac{x}{3}$$

Sin 30° equals ½, so we get $x = $ ⅔. This, in turn, allows us to determine the value of h:

$$h = \sqrt{r^2 - x^2} = \sqrt{9 - \frac{9}{4}} = 3\left(\sqrt{1 - \frac{1}{4}}\right) = 3\sqrt{\frac{3}{4}} = \frac{3}{2}\sqrt{3}$$

Inserting the value of h into the formula for the cap's surface, we get

$$C = 2\pi(3)\left(3 - \frac{3}{2}\sqrt{3}\right) = 18\pi - 9\pi\sqrt{3} = 9\pi(2 - \sqrt{3}) \approx 7.6$$

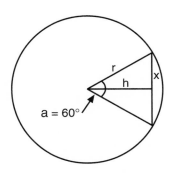

Surface of a sphere's cap

Hence the shadow that each surrounding ball throws onto the superball's surface is 7.6. On the other hand, the total surface of the superball is $4\pi r^2$. With $r = 3$ this gives 36π, or 113.1. So for how many shadows is there room on the superball's surface? Divide one number by the other, $^{113.1}/_{7.6}$, and you get 14.9.

Leech's Proof

Leech inspected the points where the 12 (or 13) outside balls come into contact with the central sphere, and—like Hoppe had done eighty years earlier—suggested weaving a net. The points of contact are the knots, and threads are woven between them. The length of each thread must be at least 1.047, to accommodate the balls.[1] In principle, the net can be woven out of triangles, quadrangles, pentagons, hexagons, heptagons, and so on.

He then proceeded to compute the minimal surfaces of each of these polygons, and added them up. Leech reaches the conclusion that the total area of the net must be greater than 0.5513, multiplied by a weighted sum of the polygons.[2] On the other hand, the net must fit tightly around the ball, so its true area must be equal to the surface of the ball. It is known that a ball with radius 1 has a surface of 4π, and so 4π must be greater than 0.5513 times the weighted sum of the polygons. Let's note that down for future reference:

$$4\pi \geq 0.5513 \text{ (weighted sum of the polygons)}$$

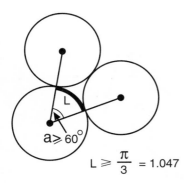

$$L \geq \frac{\pi}{3} = 1.047$$

Net, with angles at least 60° (distance between knots ≥ 1.047)

---

[1] In order for six balls to fit around the circumference of the central ball, which measures 2π, each ball requires a length of $\pi/3 = 1.047$.

[2] I will not go into the reason why this is a weighted sum rather than a straight sum. Suffice it to say that triangles carry a weight of one, quadrangles have a weight of two, pentagons a weight of three, and so on.

At this stage Euler's theorem, which we first met in chapter 5, makes another guest appearance. This time it comes in a three-dimensional guise, to provide a relationship between corners (C), faces (F), and edges (E) of a net in *space*:[3] $C + F = E + 2$. We can verify this fact for a simple net in space: a cube-like net has 8 corners, 6 faces, and 12 edges, that is, $8 + 6 = 12 + 2$. Add a pyramid to one of the cube's sides, and you get 9 corners, 9 faces, and 16 edges, that is, $9 + 9 = 16 + 2$. And so on, and so on. We can rewrite Euler's equation for nets in space as

$$2C - 4 = 2E - 2F$$

Let's note that also for future reference.

Leech now proceeded to count the edges and the faces of the net, and expressed them in terms of the polygons: Each triangle face is surrounded by three edges, each quadrangle face by four edges, and so on. Then Leech calculated the result for the right-hand side of Euler's equation, $2E - 2F$, by expressing the edges and the faces in terms of triangles, quadrangles, and so on. He did not forget to make the all-important allowance for double counting, mind you, because neighboring polygons always have edges in common. It turns out that the right-hand side of the rewritten Euler equation contains the same weighted sum of the polygons that was obtained above.

As we recall, Leech already computed a lower limit for the net's surface—0.5513 times the weighted sum of the polygons—and now he has an expression for the right-hand side of Euler's equation, which is equal to the

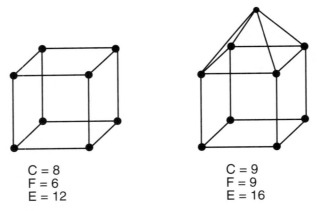

C = 8   C = 9
F = 6   F = 9
E = 12   E = 16

Euler's equation for nets noting corners (C), faces (F), and edges (E)

---

[3] That is, for polyhedra instead of polygons.

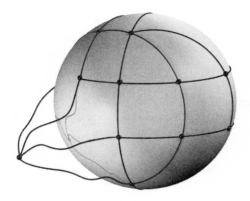

Leech's net with slack

left-hand side, $2C - 4$, in terms of the weighted sum of polygons. Putting all of this together, he saw that the area of the net must be greater than $0.5513(2C - 4)$. But since the true area is 4π, this implies that $4\pi \geq 0.5513(2C - 4)$, which, in turn, means that C, the number of knots, must be less than 14. And since each knot stands for a ball, 14 or more balls are not possible.

Wow, big deal! We never even considered the possibility of 14 or more balls touching the central sphere, the question was whether 13 could kiss or not. So, unfortunately, the first one-and-a-quarter pages of the two-page proof don't settle the Newton–Gregory dispute at all. But Leech didn't stop there: he forged ahead for another three-quarters of a page. And if you thought the fare up to this point was difficult, you ain't seen nothin' yet.

Let us assume, for argument's sake, that the net *can* be composed of 13 knots. Leech showed that this leads to an impossible situation. The first thing he did was to prove that a 13-knot net cannot have pentagons, or hexagons, or higher polygons in it because the surface of such a net would be too large. If pulled around the sphere it would sag, that is, it would have slack. So in order for the net to fit snugly around the sphere, all polygons higher than quadrangles have to be excluded. The conclusion was that the net must be woven out of triangles, and at most 1 quadrangle.

Armed with this knowledge, let's see how many triangles the net has. The clue lies in Euler's equation. With $C = 13$, that is, with 13 knots, the left-hand side of the equation, $2C - 4$, becomes 22 ($= 2 \cdot 13 - 4$). Now recall that the right-hand side of Euler's equation is equal to the weighted sum of polygons. This allows either 22 triangles and zero quadrangles,[4] or

---

[4] This is what Hoppe had used in his proof more than eighty years earlier.

one quadrangle and no, you were quite close, but the answer is not 21 triangles. For esoteric reasons that we shall not go into each quadrangle is counted twice in the sum of the polygons, and, therefore, if the net contains a quadrangle it will only have 20 triangles. Each of the two possibilities will be inspected in turn.

We first concentrate on the latter case. With a total of 21 faces (1 quadrangle and 20 triangles), the right-hand side of Euler's equation, $2E - 2F$, gives $2E - 42$. Setting this equal to the left-hand side, which was 22, and solving for E, we see that there must be 32 edges, or threads, in the net. Obviously 32 edges have 64 endpoints, and these endpoints must be attached to the 13 knots.

At this point a subtle net-weaving secret will be revealed: at most 5 threads can lead to any knot in our net. Why is that? Imagine a ball lying at, or near, the North Pole of the central sphere and threads radiating out towards 6 neighboring balls that lie around the equator. So far we have 7 balls. In order to get a total of 13 spheres, this means that 6 additional spheres would have to be added to the lower hemisphere. But there is not enough space. So let's try to make room by rolling the 6 spheres up from the equator, towards the North Pole. Well, they cannot all be rolled up without getting into each other's way. At most 5 balls can be moved northwards simultaneously. So 5 threads it will be, at most.

The details of the puzzle, that we have come up with so far, are as follows: A net must be woven out of 1 quadrangle and 20 triangles, or out of 22 triangles. It must have 13 knots, 32 threads, and 64 endpoints. Obviously, the threads must be attached to the 13 knots, and we also know that at most 5 threads can reach each knot. So let's continue from here.

We start with the case of 0 quadrangles and 22 triangles. By Euler's equation such a net must have 33 threads, and their 66 endpoints must be attached to the 13 knots. But the only way to produce such a net would be to weave 12 knots with 5 threads each, and a 13th knot with 6 threads. But that can't be done. According to the net-weaving secret, at most 5 threads can lead to any knot in the net.

Now let's inspect the case of 1 quadrangle and 20 triangles. The only way to weave 64 endpoints into 13 knots—with no more than 5 threads reaching each knot—is to attach 4 threads to 1 knot, and 5 threads to each of the other 12 knots ($4 \cdot 1 + 5 \cdot 12 = 64$). So let's weave such a net. Well we can't! And John Leech couldn't either. In fact, nobody can. It is impossible.

Is it really? In a manner rather uncharacteristic for a mathematician, John Leech decided to forego a formal demonstration of the impossibility and simply wrote that he "knows of no better proof of this than sheer trial." So let's also try. We start with the quadrangle and attach triangles to each side.

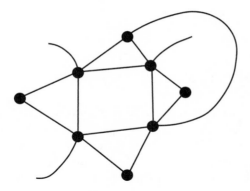

Impossible net

Then we add a free thread to each of the 4 corners of the quadrangle. Of the remaining 9 knots, 8 must have 5 threads attached to it, and 1 must have 4 threads running towards it. Try to continue the graph and note that in order to enfold the central sphere with the net, the knots and threads that lie on the periphery must be attached to one another. You will soon arrive at a situation where no more threads can be woven into knots without violating some of the details of the puzzle (20 triangles, 32 threads, 12 knots with 5 threads, one knot with 4 threads).[5]

[5] You then must also deal with the case where 5 threads lead to 3 of the 4 corners of the quadrangle, and the 4th corner has only 4 threads.

Gauss's Proof of Seeber's Conjecture

In chapter 3 we already made acquaintance with quadratic forms in two variables, which are defined as $a^2x^2 + 2bxy + c^2y^2$. Let us recall some of the important features. Geometrically, quadratic forms can be regarded as the distance between two points on a lattice. The distance between the lines that make up the lattice's grid are a in the x-direction, and c in the y-direction. The system does not need to be rectangular. The cosine of the angle between the two axes is b/ac. (Only if b equals zero, is the system rectangular.) The numerical values of x and y represent the number of ticks between the two points on each axis.

The quadratic form's determinant, D, is computed as $a^2c^2 - b^2$ (or, if negative, as $b^2 - a^2c^2$). It is called the determinant because it determines all interesting properties of the quadratic form. (Sometimes it is called the discriminant because it discriminates between quadratic forms with different properties.) For example, the surface of the lattice's fundamental cell is given by the square root of D. Two quadratic forms can have the same determinant, even if the values of a, b, and c differ.[1] From among all the quadratic forms that have the same determinant, some are regarded as being *equivalent* to each other. (They are equivalent because they can all be converted into each other by some algebraic operation that won't be described here.) Together they make up a class of equivalent quadratic forms. One member from each such class is picked to serve as its representative. It is called the *reduced* quadratic form and can be regarded, in some sense, as the simplest member from among the whole class.[2]

Lagrange, Seeber, Minkowski, and others developed algorithms to transform a quadratic form into its reduced state. We won't describe these methods here, but just ask the following question: When confronted with a quadratic form how can one tell, whether it already is reduced? As recounted in chapter 3, Joseph-Louis Lagrange gave the answer. A quadratic form is reduced *iff* (if and only if) $a \leq c$ and $2b \leq a^2$. If these two conditions are fulfilled, and only then, the quadratic form is in its reduced state.

---

[1] For example $a = 3$, $b = 4$, $c = 1.375$ and $a = 2$, $b = 5$, $c = 2.550$ have the same determinant: $D = 1.01$.
[2] All quadratic forms from a particular class can be converted into the same reduced quadratic form.

Quadratic forms are intimately related to lattices. Every lattice can be described by a base. (Bases, remember, are vectors or little arrows, one for each axis.) Usually a lattice can be described by more than one—in fact by infinitely many—different bases. The "simplest" base corresponds to the reduced quadratic form.

The quadratic forms which Lagrange dealt with are called *binary* because there were *two* variables, x and y. Seeber's contribution was to add a third variable, z, and that's what his book was about. The objects of his attention were *ternary* quadratic forms, $a^2x^2 + b^2y^2 + c^2z^2 + 2dyz + 2exz + 2fxy$. Like the binary quadratic forms, their ternary cousins have a geometric interpretation. a, b, and c are the distances between the grid lines in the x-, y-, and z-directions of a three-dimensional lattice, and d, e, and f determine the angles between the axes. Ternary quadratic forms also have determinants, which are defined as $a^2d^2 + b^2e^2 + c^2f^2 - a^2b^2c^2 - 2def$. As in the binary case, ternary quadratic forms can be equivalent and for each class of equivalent quadratic forms there is a simplest representative, the *reduced* quadratic form. How can we know whether a quadratic form is reduced? Seeber proved that a ternary quadratic form is in its reduced form, *iff* the following conditions hold:

1. $a \le b \le c$.
2. $2|d| \le b^2$, $2|e| \le a^2$, and $2|f| \le a^2$.
3. d, e, and f must be of the same sign. If they are negative,
   $-2(d + e + f) \le a^2 + b^2$.

He also proved that $a^2b^2c^2 \le 3D$ but suspected that $a^2b^2c^2 \le 2D$. When Gauss proved Seeber's conjecture he had to distinguish between two cases: either d, e, and f are all positive, or they are all negative. In the first case he introduced six new variables:

$$D = b^2 - 2d, \ E = c^2 - 2e, \ F = a^2 - 2f,$$
$$G = c^2 - 2d, \ H = a^2 - 2e, \ I = b^2 - 2f$$

Obviously a^2, b^2, and c^2 are positive. Given Seeber's conditions on reduced quadratic forms, it is easy to verify that D, E, F, G, H, and I are also positive. Gauss formed the expression $2D - a^2b^2c^2$ (the reader is invited to check the details):

$$2D - a^2b^2c^2 = a^2dD + b^2eE + c^2fF + dHI + eGI + fGH + GHI$$

All the terms on the right-hand side are positive. This means that $2D - a^2b^2c^2 \ge 0$, from which Seeber's claim follows:

$$a^2b^2c^2 \le 2D$$

Now the case must be analyzed where d, e, and f are all negative. This time Gauss introduced nine new variables:

$$J = b^2 + 2d,\ K = c^2 + 2e,\ L = a^2 + 2f$$
$$M = c^2 + 2d,\ N = a^2 + 2e,\ O = b^2 + 2f$$
$$P = b^2 + c^2 + 2d + 2e + 2f$$
$$Q = a^2 + c^2 + 2d + 2e + 2f$$
$$R = a^2 + b^2 + 2d + 2e + 2f$$

which are all positive because of Seeber's conditions on reduced quadratic forms. (Check them if you like.) Gauss then wrote down the following equation:

$$6D - 3abc = -a^2 d(J + 2P) - b^2 e(K + 2Q) - c^2 f(L + 2R) - dNO - eMO - fMN + JKL + 2MNO$$

(Check that too if you like. Otherwise you may take Gauss's word for it.) Since d, e, and f are negative or zero, all terms on the right-hand side are positive, and so is the left-hand side. Hence $6D - 3abc \geq 0$ and therefore

$$a^2 b^2 c^2 \leq 2D$$

QED

So Seeber's hunch was correct!

Apart from proving Seeber's number theoretic conjecture, Gauss also gave a geometric interpretation for this result. And that is where the implication for sphere packing lies. Consider an upright box, whose edges are a, b, and c and therefore has a volume of abc. Now reduce the box's volume by turning and twisting the axes. Try as you might, Gauss's proof tells us that the volume of the box can never be reduced by more than 29.7 percent. How come? Well this time consider a lattice with edges a, b, and c, and with a quadratic form whose determinant is D. Gauss explained that $\sqrt{D}$ is the volume of the lattice's fundamental cell (the "box"). So the theorem conjectured by Seeber is equivalent to saying that no grid system exists whose fundamental cell is smaller than the volume of the upright box, reduced by 29.3 percent:

$$\sqrt{D} \geq abc/\sqrt{2} = 0.707\,abc = (1 - 0.293)abc$$

Does that mean that a box can never be squashed? That you can never reduce its volume to zero by stepping on it? Well, not quite. Gauss's statement holds only for boxes that are in *reduced* form. This means that the values of *d, e,* and *f,* which determine the angles of the box, must fulfill the three conditions mentioned above. A squashed box is *not* equivalent to the upright box, even though it may have the same edge lengths *a, b,* and *c.* So for all you recycling enthusiasts: Go on, you can do it! Squash the box.

In order to be a contender for the best packing, a lattice's box must be able to fit one ball of radius 1 (and of volume $\frac{4}{3}\pi = 4.189$) at each corner. So the edges must have lengths of at least 2. Hence, the upright box has a volume of $2 \cdot 2 \cdot 2 = 8$. On the other hand, we know that the volume of a box can not be reduced by more than 29.3 percent. Therefore, the smallest possible box has a volume of 5.657. From this it follows that the best packing density is $\frac{4.189}{5.657} = 74.05$ percent.

Which box has this minimal volume? The question is equivalent to asking under what conditions $a^2b^2c^2$ is equal to $2D$. For this to hold, the right-hand sides of the two above equations must be equal to zero. We will just skim the surface here. Gauss said that the *cosines* of the grid's angles are d/bc, e/ac, and f/ab. The right-hand sides of the equations are equal to zero *iff* these three ratios are equal to $\frac{1}{2}$. And the angle for which the cosine is $\frac{1}{2}$ is 60°. And a box whose edges are inclined at 60° angles represents just the FCC and the HCP arrangements. Now how 'bout that?

Density of Spheres in a Dodecahedron

First, the dodecahedron is cut into twelve pyramids. They all have pentagonal bases. Let us denote the length of the edges of the pentagons with the letter *a*. That was the easy part. From here on things become a bit complicated. Actually, all that is needed is just elementary mathematics (Pythagoras and some trigonometry). But the derivations are long and tedious. In fact, they are so convoluted that for the most part I will just state the results.

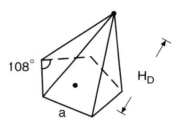

Volume of a dodecahedron

1. The volume of a pyramid is calculated as

$$V_D = \frac{1}{3} \, (\textit{Surface of the pentagonal base}) \cdot \textit{Height}$$

2. The surface of the pentagonal base is

$$S_D = 5 \, (\tan 54°)\left(\frac{a}{2}\right)^2 = \frac{5}{4} \, (\tan 54°)a^2$$

The "54°" arises from the fact that the angle of a regular pentagon is 108°.

3. The height of the pyramid is

$$H_D = \frac{1}{2} \, (\tan 54°) \tan \left(\arcsin \left(\frac{1}{2 \sin 36°} \right) \right) a$$

4. The volume of the dodecahedron is

$$V_D = 12 \cdot \textit{Volume of the pyramid}$$

$$= 12 \cdot S_D \cdot H_D \cdot \frac{1}{3}$$

$$= \frac{5}{2} (\tan 54°)^2 \tan \left(\arcsin \left(\frac{1}{2 \sin 36°} \right) \right) a^3$$

$$\approx 7.66a^3$$

5. The radius of an inscribed sphere is

$$r = \frac{a}{20} \sqrt{250 + 110\sqrt{5}}$$

Hence the edge length of the dodecahedron with an inscribed sphere of radius 1, is

$$a = \frac{20r}{\sqrt{250 + 110\sqrt{5}}} = 0.898$$

Therefore, the volume of the dodecahedron is $V_D = 5.55$.

6. Since the volume of a sphere is 4.1888 (= $\frac{4}{3}\pi$), the density is $^{4.1888}/_{5.55} =$ 75.47 percent.

Density of Spheres in a Tetrahedron

Refer to the figure!

1. Volume of a tetrahedron with edge–length 2:
   The surface of the base triangle is, by Pythagoras,

$$S_T = \sqrt{3}$$

The height of the tetrahedron, H_T, is computed from (again by Pythagoras)

$$H_T^2 + X^2 = 2^2$$

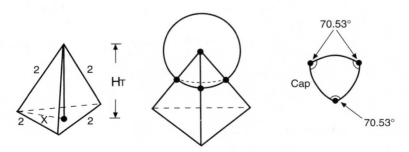

Density of spheres in a tetrahedron

We know that $X = 2/\sqrt{3}$ (just substitute 2 for D in the first section of the appendix to chapter 4). Hence

$$H_T^2 + \frac{4}{3} = 4$$

and from this it follows that $H_T = \sqrt{\%}$.
The volume of a pyramid (and hence of the tetrahedron) is:

$$V_T = \frac{1}{3} \ (\textit{Surface of base triangle}) \cdot \textit{Height}$$

$$= \frac{1}{3} \ \sqrt{3} \sqrt{\frac{8}{3}} = \frac{\sqrt{8}}{3} = 0.943$$

2. Volume of the parts of the sphere contained within the tetrahedron:
    If a ball is placed at the corner of a tetrahedron, a spherical triangle is cut out from it. We shall call it a "cap."
    Spherical trigonometry tells us that the surface of the cap with N corners is the sum of the angles, minus $(N - 2)\pi$.
    For a spherical triangle, that is, for a cap that has three corners, N equals 3. What are the angles of the cap?
    Start by drawing lines from the middle of an edge to the two opposite vertices. By Pythagoras's Theorem, these lines have length $\sqrt{3}$.
    The *sine* of half of the angle that lies between these lines is $1/\sqrt{3}$. Hence the angle is 70.53°. The surface of the cap is, therefore, $3 \cdot 70.53° - \pi = 31.6°$.
    On the other hand, the total surface of a sphere is $4 \cdot \pi = 720°$. Hence 31.6° represents 4.389% of the total surface.
    The *volume* of the part of the sphere that is contained in the tetrahedron is proportional to its *surface*.
    Therefore, the balls at the four corners represent 17.55% of a sphere's volume.
    The volume of a sphere is $\frac{4}{3}\pi = 4.189$, and 17.55% of 4.189 equals 0.735.
3. The density is, therefore, $^{0.735}\!/_{0.943} \approx 77.97\%$.

Density of Spheres in an Octahedron

We investigate a square pyramid, which is half an octahedron.

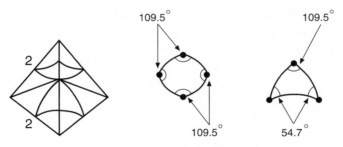

Density of an octahedron

1. Volume of the pyramid with edge-length 2:
    The surface of the base quadrangle is $2 \cdot 2 = 4$.
    The height of the pyramid is $\sqrt{2}$.
    The volume of the pyramid is

$$V_o = \frac{1}{3} \text{ (Surface of the base quadrangle)} \cdot \text{Height}$$

$$= \frac{1}{3} \, 4\sqrt{2} = 1.885$$

which is exactly twice the volume of the tetrahedron.

2. Volume of the parts of the sphere contained within the pyramid:
    There is one ball centered at the top of the pyramid, and there are
    four balls centered at the four corners of the base.

(a) The ball at the top cuts out four angles of 109.5° each. Again, this requires no more than Pythagoras's theorem and trigonometry, but we won't go into the details. The surface of this four-cornered cap is

$$S_1 = \text{(Sum of the angles)} - (N - 2)\pi$$

where N, the number of corners of the base, is 4 in this case. Hence the surface contained within the pyramid is

$$S_1 = 4 \cdot 109.5° - 360° = 78°$$

This corresponds to 10.8 percent of a whole sphere.

(b) The four balls at the bottom corners cut out triangles. Without going into details, the spherical triangles have one angle of 109.5°, and two angles of 54.7°. Therefore the surface of each ball contained within the pyramid is (this time N equals 3)

$$S_2 = 109.5° + 2 \cdot 54.7° - 180° = 38.9°$$

This corresponds to 5.4 percent of a whole ball.

(c) To summarize, there are four balls at the corners of the base, and one ball on top. Hence, the total volume of those parts of the five balls that are contained within the pyramid is

$$S = 4S_1 + S_2$$
$$= 4 \cdot 5.4\% + 10.8\%$$
$$= 32.4\% \text{ of a whole ball}$$

Since the volume of a whole ball is 4.189 ($= \frac{4}{3}\pi$), the volume of the parts of the spheres contained in the pyramid is 1.357.

3. Dividing one volume by the other, we get

$$Density = \frac{1.357}{1.885} \approx 72\%$$

Number of Triangles in the Net

X triangles on their own have $3X$ edges. Since the triangles are woven into a net, each thread belongs to two triangles at the same time. Hence there are only $\frac{1}{2}(3X) = 1.5X$ threads in the net. Euler's formula, which we discussed in chapter 6, says

$$\text{Vertices} - \text{Edges} + \text{Faces} = 2$$

or

$$\text{Faces} = \text{Edges} - \text{Vertices} + 2$$

In other words,

$$\text{Triangles} = \text{Threads} - \text{Neighboring Spheres} + 2$$

With X triangles, $1.5X$ threads and 44 neighboring spheres we get

$$X = 1.5X - 44 + 2$$

Which gives

$$X = 84$$

Hence there are 84 triangles in the net.

Number of Neighboring Spheres

On the other hand, Hales showed that the nets of Delaunay stars could be made up of at most 102 threads. Since each thread belongs to 2 triangles, the net has at most 51 triangles. Euler's formula

$$\text{Triangles} = \text{Threads} - \text{Neighboring Spheres} + 2$$

translates as

$$51 = 102 - \text{Neighboring Spheres} + 2$$

Hence, the number of Neighboring Spheres is at most 53.

The Proof—An Explanation

For his proof, Tom Hales proposed three things: weave a net, partition space, and partition space another time.

To begin, think about a saturated packing—a collection of balls that is packed so tightly that no additional balls can be added. Pick one sphere at random and call it the nucleus. The net is woven in the following manner: draw red lines from the center of the nucleus to the centers of the neighboring spheres.[1] This produces a sort of wire mesh. Mark the points where the wires cut through the nucleus. The nucleus now sports a set of freckles. Connect the neighboring freckles along the surface of the nucleus with yellow threads. This defines a net that is wound around the nucleus. It is made up of loops of various shapes.[2]

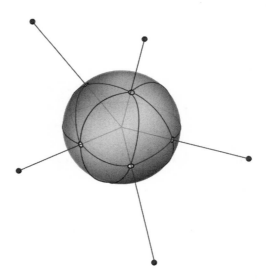

Wire mesh and net around
the nucleus

[1] We call spheres "neighbors" if their centers lie at a distance of no more than 2.51 from each other. The colors (red, yellow, and so forth) are mentioned only for illustrative purposes. Our pictures are just black and white.
[2] The word "loop" usually conveys roundness, but this need not be so. A net's loops have corners, that is, they are polygons.

Next, Hales suggested spanning the wires with red walls. These walls define a set of red tetrahedra. Finally, he also proposed forming Voronoi cells around the nucleus and around the neighboring spheres. The walls of the cells were to be painted blue. Space has thus been partitioned twice, once into red tetrahedra and another time into blue cells.

Now star-like structures are built around each sphere. Its components are the red tetrahedra and the blue cells. If a red tetrahedron is sufficiently small it is used as a building block, otherwise parts of the corresponding blue cell are selected.[3] Once the star has been built, it resembles a blue cell, with inverted red tetrahedra sticking out from some of its sides. This multicolored star represents Hales's hybrid approach to Kepler's conjecture.

Nets and stars are interchangeable. The threads of the net, you see, are exactly the lines where the star cuts through the shell of the nucleus, and that makes for a close relationship between stars and nets. Thus, by defining and measuring the net's score, the score of its star is also determined. But all these yellow threads make for a rather drab net. Let's put some life into it by coloring all loops where red parts of the star cut through the shell in orange (yellow + red = orange), and all loops where blue parts of the star cut through the shell in green (yellow + blue = green). When a green loop lies next to an orange loop, the threads will be orange on one side and green on the other.

The remainder of the proof consists of computing the scores of all possible nets. As Hales soon came to realize, a net's score is determined precisely by the shapes of the loops and by the colors of the threads. Apparently, even in the most egalitarian of all worlds, shape and color do matter. Color determines the mode of measurement: orange-colored loops are scored by their *surplus,* green loops are scored by their *premium.*[4] (For threads that are colored orange on one side and green on the other, the appropriate method is used for each of the neighboring loops.) The only remaining problem are the shapes of the nets' loops. All told, Hales had managed to reduce the packing problem to the contemplation of stars, the contemplation of stars to the inspection of nets, and the inspection of nets to the scrutiny of loops. It was definitely getting simpler and simpler.

As was pointed out in the previous chapter, a score of 8 pts corresponds to a density of 74.05 percent. The basic philosophy behind Hales's proof

---

[3] By "sufficiently small" Hales designated tetrahedra whose circumradius, that is, the radius of the smallest ball that contains the tetrahedron, is at most 1.41. (The regular tetrahedron, with edge-length 2, has a circumradius of 1.22.)

[4] The notions of "surplus" and "premium" were introduced in chapter 11.

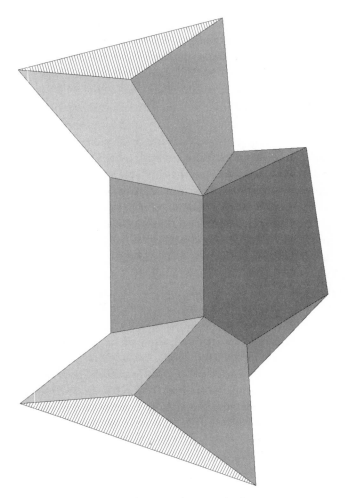

Cell with protruding tetrahedra

was to show that regular triangles score 1 pt, other triangles score between 0 and 1 pt, four-sided loops score 0, and all other loops give negative scores. Hales conjectured, indeed he was convinced, that when the scores of all loops are added up, no net could ever achieve more than 8 pts. And only two nets could achieve a score of exactly 8 pts. Which sphere arrangements do these nets represent and why do they score exactly 8 pts?

The answer to the first question is: Kepler's sphere arrangements. Both the FCC and the HCP consist of eight regular tetrahedra and six regular octahedra. Hence, the corresponding nets consist of eight equilateral triangles and six squares. What are the scores of these nets? *Surplus* and *premium* were designed in such a way that regular tetrahedra, and hence triangles, score 1 pt, regular octahedra, and hence squares, score 0 pts. Hence the eight triangles score a total of 8 pts. The six squares, on the other hand, add nothing to the total score. Therefore the nets, and hence the stars, and hence the sphere packings of Kepler's arrangements score exactly 8 pts.

The assignment that Hales faced was to show that no other net could also score 8 pts. Hales laid out a game plan. First he would classify all nets into a few distinct groups according to the shapes of the loops. Then he would seek the representative net with the maximum score within each group. Finally he would show that this maximum score always lies below 8 pts. Should he be successful with all steps, he would have shown that all nets, without exception, score less than 8 pts. That would be tantamount to saying that no packing—except for Kepler's arrangements—could reach a density of 74.05 percent. But he was not going to complete his game plan overnight. It would take four years of intensive work to eliminate all conceivable nets.

In Segment 1 Hales was going to show that all the nets that are composed entirely of triangles have scores of at most 8 pts. He began his work by determining the characteristics that are necessary so that a net could at least provide the hope of achieving a score higher than 8 pts. How many knots must it have? How many threads may lead to a knot? How long must the threads be? To investigate these questions, Hales derived relationships between the various attributes of the net's triangles. Some of them were very esoteric. One of the simpler relationships was "the triangle's score is less than 0.5 pts if one of the threads is longer than 2.2 but shorter than 2.51."[5] More complicated examples were "the score is smaller than 0.287389, minus 0.37642101 multiplied by the angle of the corresponding tetrahedron." Or "if the sum of the three threads is less than 6.3 and the triangle cannot be contained within a circle of radius 1.41, then the angle is greater than 0.767."

Altogether Hales required thirty-five such inequalities. He could have proven them with pencil and paper but that would have been very boring. Moreover, he suspected that hundreds of such statements would turn up during the coming years. It would be so much faster if he had a computer program that would automate this tedious task. Designing a program that automatically proves relationships and inequalities between variables also

---

[5] To be exact, the measurements do not relate to the lengths of the threads. Rather they refer to the lengths of the edges of the Delaunay stars. The threads are the projections of these edges onto the central ball and are, therefore, shorter.

takes a lot of effort. However, Hales decided that in the long run, it would be more efficient to write a program and get the whole thing over with, once and for all.

Say you have two functions and you want to show that one of them is always smaller than the other one. Imagine them as curves on a sheet of paper. Subdivide the sheet into squares, and check whether the squares that the first curve enters always lie below the squares that the second curve enters. If they do, you are done and the inequality has been proven. But if the two curves overlap in some of the squares, the offending squares are divided into subsquares and the checks are redone with a magnifying glass. With any luck, you will be done at this stage. But if there still remain offending squares, redivide them into subsubsquares, use a microscope, and so on. If the inequality is, in fact, true, then you will eventually find that all minisquares that the first curve enters lie below the minisquares of the second curve. (If after several subdivisions of the boxes the inequality still cannot be established, then it may simply not be true.)

This is how it's done in two dimensions. In higher dimensions the work becomes more tedious but not more difficult. Simply replace the squares by n-dimensional boxes and apply the same process. Hales let the computer do the dirty work for all thirty-five inequalities. One of them required just seven boxes before it was proven to be correct, another inequality required over two million. Eventually the truth of all statements was confirmed.

Based on the thirty-five inequalities, Hales realized that for a net to approach the score of 8 pts it would have to satisfy nine requirements. For example, it has to be composed of between 13 and 15 knots, and 4, 5, or 6 triangles must meet at each knot. The net can not contain more than 2 knots where 4 threads meet, and these 2 knots can not lie next to each other. And so on. Obviously such requirements significantly cut down on the number of nets that could possibly score higher than 8 pts. Hales suspected that maybe several hundreds of them would exist. He set out to use

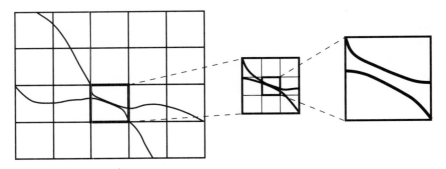

Proving inequalities through magnification

combinatorial techniques to generate a list of the nets and to inspect them one by one.

At first he thought of having the computer generate the list of all nets that fulfill the nine requirements. But here we have an example of where the human mind is still superior to electronic computers. The human who accomplished this feat was none other than Doug Muder, the world record holder for the lowest upper bound. Without the use of a computer, just with pencil and paper, he gave a direct classification of all nets that satisfy the nine requirements. By that time Muder had become somewhat of a mathematical recluse. His disenchantment with academia was so intense that he did not even bother publishing his result. He simply asked Hales to append it to one of his articles.

Based on Muder's result, Hales was able to show which nets fulfill the nine requirements. There were not hundreds of them, as he had initially suspected. There were not even dozens of them. No, there was exactly *one* net that fulfills all nine requirements. It was composed of twenty-four triangles connected through fourteen knots. The proof that it scores less than 8 pts turned out to be a piece of cake. All Hales had to do was reuse some of the thirty-five inequalities.

There was no time for rest, however: on to Segment 2, which dealt with *n*-sided loops, where *n* is three or greater.[6] Hales was going to show that three-sided loops score at most 1 pt, rectangles score at most zero points, and any loop with more than four sides gets a negative score. In other words, all loops except triangles and the square waste so much space that a penalty must be imposed on the net's total score! Hence, high-scoring nets should include as many triangles as possible. If differently shaped loops are needed, preference should be given to squares. Any other shape should be avoided if at all possible. If they must be included, they should be accompanied by additional triangles to compensate for the penalties.

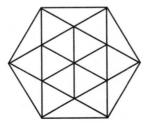 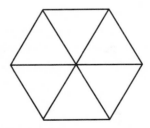

Muder's net (Hales's diagram 6.2)

---

[6] The results of this segment were not needed immediately, but prepared the ground for later parts of the proof.

This is the point where the results of Hales's dreams first bore fruit. Up until now, he had always computed the scores of stars by their *surplus,* that is, by the density gain (or handicap) they have over octahedra. The method had been quite an advance over previous efforts, since all attempts to use Voronoi cells had ended in serious problems—as Fejes-Tóth's faulty dodecahedral conjecture had shown beyond doubt.[7] So the idea with stars was a step forward. But it was by no means the answer to all problems. While it gave better bounds on the packing density for small simplices, it ran into trouble whenever the bits and pieces of the star became too large. Specifically, when the size of the tetraehdron reached about 1.8, the surplus estimates were totally off the mark.[8]

So one method had led into a blind alley, the other to a dead-end road. Was there a way out? There was! Combine the blind alley with the dead-end road, knock out the obstructing wall, and blaze a new trail. This is what Hales did. His brainstorm produced a hybrid scoring method that combined density estimates of both simplices and cells. Take a tetrahedron with a sphere placed at each of its four corners and an octahedron with a sphere at each of its eight corners. The density surplus of a tetrahedron was defined as its gain or handicap over the octahedron's density (72.09 percent). The *premium* is defined similarly. Take a ball's Voronoi cell and slice it into wedges. Each wedge contains, at its lower corner, part of the ball. Compute the density that prevails in the wedge and compare it to the density of the octahedron. The gain or handicap is the wedge's *premium.*

Hales incorporated the *surplus* of the tetrahedra and the *premium* of the cells into the *score.* If the size of a tetrahedron was smaller than 1.41, its score was computed by the surplus, as in Segment 1. But whenever the tetrahedron's size was larger than 1.41, Hales would switch to the computation of the *premium* of the appropriate cell.[9] The combination of the two methods of measurement produces a useful score for stars of all sizes. This hybrid approach retains the best features of both methods, without producing any harmful side effects. Hales had some additional fancy ideas—like reassigning offending corners of the Voronoi cells to its neighboring cells—but we won't burden ourselves with those details.

The stage was set for the inspection of the loops. Hales already knew that

---

[7] Recall, that Fejes-Tóth showed that the dodecahedron is the smallest possible cell. But since dodecahedra do not tile space, they have to be combined with other shapes to form an efficient packing.

[8] Again, by "size" of a simplex Hales meant its circumradius.

[9] The cutoff number of 1.41 was a bit arbitrary. All that mattered was that it was safely below 1.8.

regular triangles score exactly 1 pt. In addition, one of the thirty-five inequalities that Hales's computer program had spat out for Segment 1 implied that irregular triangles score between 0 and 1 pt. To show that loops of any other shape score at most 0 pts, Hales returned to the star, which gave rise to the net. He split it into a number of wedges, each of which was shaped either as a small tetrahedron or as one of four different archetypes. He was going to show that the density of each wedge was at most 72.09 percent. Since this number represents the density of the octahedron, the score would at most be zero.

Hales went to work. Wedges were broken into subwedges, the computer was programmed to spit out some more inequalities, and the surpluses and the premiums of all the bits and pieces were computed. One by one, each of the four archetypes was shown to have a density less than the octahedron. This implied that loops of any shape other than triangles or squares have negative scores. A second implication was that whatever pieces are needed to fill the space around a tetrahedron, their densities could never be greater than that of the octahedron.

With Segment 1 and Segment 2 safely under his belt, Hales's next challenge was Segment 3. This is where he dealt with nets that are woven out of triangles and quadrangles. He was going to prove that no such net could have a score higher than 8 pts. Of course, Kepler's arrangements also consist of triangles and quadrangles, and we know that they score exactly 8 pts. The question was whether another net could also reach 8 pts. Hales thought that this was out of the question.

First let us look at squashed versions of Kepler's arrangements. Can a deformation of the net raise its score? The answer is no. Even the slightest deformation transforms regular triangles into irregular ones, thus reducing the score to below 1 pt, and deforms squares into rectangles, making their scores negative. Hence, the net's total score falls to below 8 pts and squashed versions of Kepler's arrangement were no contenders for the perfect score. What about other nets composed of triangles and quadrangles? Maybe a few triangles, which have positive scores, could be added to an existing net, thus increasing its score. Hales was convinced that this was impossible. Or so he thought. As he was to discover soon there was one net, consisting of ten triangles and five quadrangles, which came pretty close to the perfect score. It was the dirty dozen. Hales did a wise thing. Instead of letting himself be frustrated by this annoying net, he looked it straight in the eye—and ignored it. The dirty dozen, which requires very delicate treatment, would be dealt with in a separate segment.

But we are jumping ahead. Hales started his work by laying out a plan similar to the one that had been successful for Segment 1. First, he was going to establish a list of properties that are necessary in order for a net to have a hope of reaching a score higher than 8 pts. Then he was going to list all nets that satisfy these characteristics, and sort them into classes. Following that, he would seek the maximum score in each of these classes. Hales hoped he would be able to show that these maxima always lie safely below 8 pts. Thus he would eliminate all nets, one by one. But it was a frightening prospect. In Segment 1 he had been very lucky because only a single net existed that had a chance of scoring high. It had easily been eliminated. In Segment 3 there could be millions and millions of them, in which case the task would be hopeless. Would he be lucky again?

The search for the required properties was based on twenty-six inequalities between the attributes of loops and their scores. It turned out that a net woven out of triangular and quadrangular loops had to satisfy exactly seven requirements in order to even have a prayer of scoring 8 pts or more. For example, acceptable nets had to include at least eight triangles and at most six quadrangles. Knots were not allowed to connect one triangle to four quadrangles. And so on.

Generating a list of those nets was one of the most important parts of Segment 3. In the search for all nets that satisfy the seven requirements the computer was put to work again. The algorithm starts out with partially completed nets. Additional loops are added by tying loose ends into knots. Existing nets are modified by adding threads and creating subloops. When all loose threads have been tied up, the program performs a quality check: Are the seven requirements fulfilled? If they aren't, the net is discarded. Only finished nets that *do* satisfy the seven requirements are contenders for a high score and are put on the list of candidates. After hours of whirring and buzzing the computer coughed up the result: 1,749 nets. At first, Hales was taken aback. This certainly was a lot of nets. But then he breathed a sigh of relief; the problem was not unmanageable. The important point was that an explicit list had been obtained.

Now it was time to verify whether the scores of the 1,749 nets lie safely below 8 pts. The computer spat out fifty-one inequalities, which related the scores of the loops to the lengths of the threads and to the angles at the corners. Additionally, there was constraint number fifty-two. It was actually an equation that stated the obvious fact that the sum of the angles around a knot must add up to 360°. Using these fifty-two constraints, the computer was ready to compute the maximum scores that the nets could attain. Would they lie below 8 pts?

Unfortunately, it is notoriously difficult to maximize non-linear equations in many variables, especially if the number of triangles, quadrangles, and knots must be integers. But once again, Goddess Fortuna was kind to Hales. In all but twenty-three of the 1,749 cases the non-linearities could be avoided by employing a technique called "linear relaxation." The technique consists of simply dropping the integer requirement and some of the non-linearities. Let us describe the technique in a few words.

The fewer constraints there are in a system, the more freedom exists. This is as true in a mathematical system as it is in a political system. Under a dictatorship, for instance, with its numerous restrictions on all forms of human activity, possibilities for success in most endeavors are severely restricted. As a consequence the economy is inefficient, quality of life is low, and art is boring. In a free country, on the other hand, where everything is more relaxed and everyone is allowed to do as one pleases, innovation flourishes and the economy booms. Dropping or relaxing constraints in a maximization problem has an effect similar to the liberation of a country from a dictatorship. Suddenly there is more freedom, and with more liberty to move around, the solution will in all likelihood be better. The upper bound of a maximization problem certainly won't be lower, and most probably will be higher. So if the linear relaxation, which can be found with relative ease, were to produce a score that lies safely below 8 pts, this would be all the more true for the nonlinear integer problem. And this, after all, is what Hales wanted to prove. The actual technique he used to find the linear relaxed maxima is called the "simplex method." It is one of the most important computer algorithms of the twentieth century.[10]

In 1,726 of the 1,749 cases, the linear relaxed version of the problem gave scores below 8 pts. There remained twenty-three nets—four of them obvious, and nineteen pathological cases—where linear relaxation produced bounds of 8 pts or above. Of the obvious nets, two belong to Kepler's packings, of which it had been known all along that they would score exactly 8 pts. Then there was the icosahedral net. This net corresponds to the dodecahedral Voronoi cell. And the dodecahedral Voronoi cell, which contains only triangles, had been eliminated in Segment 1. Finally there was the dirty dozen, which Hales decided to ignore for the time being.

The real problem now was the nineteen remaining nets. Their number diminished to eighteen after Hales realized that two of the nets were actually mirror images of each other. The fact that these pathological nets could not be eliminated did not mean that their scores actually reached or sur-

---

[10] This method, as well as the "branch and bound method" mentioned on the following page are described in a little more detail in chapter 12.

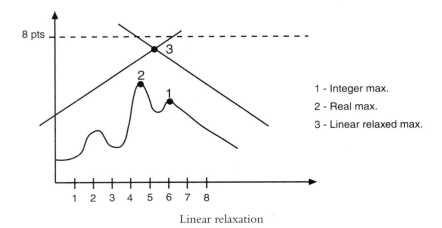

1 - Integer max.
2 - Real max.
3 - Linear relaxed max.

Linear relaxation

passed 8 pts. All it meant was that the methods that Hales had used were not powerful enough to establish that the true scores were, in fact, lower than 8 pts. Hence, these cases required more refined treatment.

Hales started his work on the pathological nets by tuning up the linear programs. He divided the domains of the optimization problem into several thousand smaller sets and worked on them. Eventually he was able to show that fourteen nets have scores below 8 pts, and they could be eliminated. Two additional nets were dealt with by application of an advanced optimization technique called the B&B method. Only B&B does not stand for "bed and breakfast" but for the "branch and bound" method. These two holdouts could not withstand the double barrage of enhanced linear programs and B&B. The upper bound for their scores were pushed below 8 pts.

The last two nets required even more specialized treatment. Upon closer inspection it turned out that the wedges of the star that gave rise to the offending nets were rather small. This allowed the addition of extra constraints. With more restrictions, elbow room became cramped. And with less freedom to move around, the upper bound certainly can't be higher. Most probably it will be lower. (Actually, the addition of the extra constraints counteracted the "linear relaxation": when the integer-requirement was dropped, the virtual nets were free to reach a higher score.[11] Subsequent efforts then concentrated on pushing the relaxed bounds back down, through the addition of new constraints.) And this was exactly what hap-

---

[11] *Virtual* because a real net can't contain, say, 2.5 triangles.

pened. As soon as Hales added the extra constraints, the bound dropped below 8 pts. The last two nets went out the window.

And that was it: 1,745 nets with triangular and quadrangular loops had been eliminated. The only nets whose scores could reach or surpass 8 pts belonged to the dirty dozen, to be dealt with in Segment 5, to the icosahedral net, dealt with in Segment 1, and to Kepler's two packings.

Next on the list was Segment 4. It dealt with nets that contain loops other than triangles and quadrangles. Hales argued that loops with more than four sides take up too much space, give too little in return, and have such incompatible shapes that they cannot be part of a winning strategy. Just one such loop would make the whole net incompatible with a high score. No amount of triangles would be able to make up for the squandered space. To make his point, Hales wrote a forty-seven-page paper, densely filled with equations and formulas. But even forty-seven pages did not suffice and in order to present all the evidence he had to write an additional sixty-three-page paper.

The procedure was similar to the one used in Segments 1 and 3. First, the characteristics that are necessary for a net to obtain a score of 8 pts had to be found. Then, using combinatorics, a list of all nets that fulfill these characteristics had to be generated. Finally, it had to be ascertained that the scores of all nets on the list do not, in fact, reach 8 pts.

Some of the required characteristics are, for example that the net has at most 100 knots, that two knots with four threads each cannot lie next to each other, and so on. One important fact is that nets that aspire to a score of 8 pts cannot include nonagons (a polygon with *nine* sides). Neither can they include decagons, or any polygon with more than ten sides. The reason is that such loops waste too much and add too little. On balance, they make a negative contribution to the net's total score.

By the way, there is a maximum score that can be attained. No matter what its shape, no net can ever score more than 22.8 pts. This significant fact is implied by the upper bound on the packing density that Rogers had discovered in 1958 (77.97 percent, as shown in chapter 9).[12] It corresponds to four balls placed at the corners of the tetrahedron and would be the maximum packing density if tetrahedra were able to fill space without gaps. Since they aren't able to do so, 77.96 percent or 22.8 pts are the maximum density and score that a star or net can achieve locally.

To make use of this interesting fact and to complement the definition of the score, Hales introduced an additional measure for stars and nets: the amount squandered. While the score is the improvement in density over the

---

[12] 22.8 pts = $22.8 \times 0.05537 = 1.2624 = 77.96\%$.

octahedron's 72.09 percent, the amount squandered is the deficit as compared to the tetrahedron's 77.96 percent. The fact that no net can have more than 22.8 pts has a very important consequence: if a loop, or a collection of loops, squanders more than 14.8 pts, the net must score less than 8 pts. (If more than 14.8 are deducted from 22.8, the remainder is less than 8.)

Once again it was time to utilize combinatorial techniques to generate a list of nets with the required characteristics. The computer whirred and buzzed. The printer clanked and clunked. Finally Hales held the output in his hands: there were 2,469 nets that contained at least one pentagon, 429 with at least one hexagon, 413 with at least one heptagon, and 44 nets that included at least one octagon. Altogether 3,345 nets that had the potential to reach, or even to surpass, a score of 8 pts were listed. They were subjected to the by-now-familiar linear program. The maximum score was sought, given constraints on the angles, thread lengths, placement of loops, you name it. As the nets were inspected loop by loop, the computer kept track of how much space was being squandered. As soon as the squandered space reached 14.8 pts, the net was thrown onto the pile of rejects. No amount of space-saving triangles would have been able to make up for the lost space.

The method worked in most cases: 3,156 nets could be chucked out immediately. But 189 nets did not succumb to the attack. Linear programming methods were not able to prove that the scores were below 8 pts. These nets required special attention and Hales had to call in heavier artillery.

Taking his cue again from a good dictatorship, Hales restricted the nets' freedom by adding more constraints. Thus he prevented them from attaining a high score. The new constraints were by no means simple. Number 131574415, for example, reads: "The score is smaller than 1.01 minus 0.1 times the length of thread 1 minus 0.05 times the length of thread 2 minus 0.05 times the length of thread 3 minus 0.15 times the length of thread 5 minus 0.15 times the length of thread 6, provided that the angle is less than 1.9, thread 1 is shorter than 2.2 and thread 4 is longer than 2.83." Or take constraint 941700528: "The space squandered is greater than 0.14 times the length of thread 5 plus 0.19 times the length of thread 6 minus 0.676, provided that threads 1 and 4 are equal to 2, thread 5 is longer than 2.83 but shorter than 3.23, and thread 6 is longer than 3.06 but shorter than 3.105."

Such fierce-looking inequalities really took the juice out of the majority of the remaining 189 nets. One hundred and one were defeated by the augmented linear program. But eighty-eight obstinate, stubborn, tenacious nets still resisted all efforts. They had to be inspected and treated case by case. The offending loops were divided into smaller subloops, and B&B was applied. Thus the eighty-eight holdouts were also dealt death blows and Segment 4 was brought to an end.

Now all that was missing to complete the proof was the dirty dozen. Hales's early computer experiments had produced disturbingly high results for this configuration. Subsequently he decided to ignore the wayward configuration while getting the rest of the proof in order. But now there were no more excuses. The troublesome configuration had to be met face on. The dirty dozen was the task of Segment 5.

Let's recall that Hales usually did not compute the score itself, but only upper bounds for the true, unknown score. The best upper bound that he had found for the dirty dozen so far was 8.156 pts. That didn't help much. The dirty dozen's score lies below 8.156, but is it below 8.000? Only if this question is answered to the affirmative can Kepler's conjecture be considered proven. If Hales and Ferguson could not reduce the bound below 8 pts, all would be lost. While there would be no certainty that the dirty dozen scores 8 pts or above, there would always remain a doubt. Hales put Ferguson to work.

The dirty dozen's net is made up ten triangles and five quadrangles. In principle, it should have been treated in Segment 3 where nets with such shapes were analyzed. But while working on that segment, it soon become apparent—indeed, Hales had suspected as much from his earlier computer experiments—that the necessary estimates would be much more delicate than for the other triangular nets.

With the help of their trusted friend, the computer, Ferguson proved various relationships between the scores and the angles of the two kinds of loops. Then he multiplied the inequalities for the triangles' score by ten, and the inequalities for the quadrangles' score by five. Finally, everything was added up. During the computations two facts came in handy: the sum of all angles that meet at a knot is $360°$ (= 2π), and the sum of all solid angles that meet at the center of the nucleus is 4π. After many trials and tribulations, Ferguson received the following equation for the dirty dozen's total score:

$$\text{Total score of the dirty dozen} \leq 5b + 10c - 4\pi m$$

where $b = 0.49246$, $c = 0.253095$, and $m = 0.3621$. Substitute these numerical values into the equation and there you have it: The right-hand side of the equation comes out to 7.99961! That's an upper limit for the net's score. The derivation of this limit may seem rather simple and quite straightforward, but nothing could be further from the truth. It was an extremely difficult, arduous task, that required the solution of many complex problems on the way.

Hence, the dirty dozen or, to return to its scientific name, the pentagonal prism, scores less than 7.99961 pts. Ferguson could have done even better than that and proved a tighter bound, like 7.98 pts, but why bother? It was quite sufficient to show that the dirty dozen's score was below 8 pts, and that has been done with 0.00039 to spare.

With this the last segment of the master plan had fallen into place. The pentagonal prism that created such problems in the earlier papers was no longer a contender for the "best packing" title. Hales and Ferguson breathed a collective sigh of relief: Only the two sphere arrangements that Johannes Kepler had described four hundred years earlier—a dozen spheres appropriately arranged around the central sphere—score a perfect 8 pts. They are the *arctissima coaptatio,* the densest packing. The proof was complete. Kepler's conjecture had finally been solved.

QED

Altogether Hales had checked the scores of 5,093 nets, and Ferguson had checked one additional net. (It's not as if Hales had done 5,093 times the work, it's just that Ferguson had been saddled with the toughest, most vexatious specimen of them all.) The vast majority of the nets were eliminated by the computer. About 100 of them had to be checked by hand with more refined methods. One by one, they too were eliminated. All, that is, except for the FCC and the HCP. By the way, let us recall that Barlow had shown in 1907 that there are an *infinite* number of arrangements—albeit not lattice arrangements—that achieve a density of 74.05 percent (see chapter 1). Upon close inspection of those arrangements, we see, however, that they are all made up of FCCs and HCPs.

So the FCC and the HCP really are the densest packings. Surprised? Hardly. Most mathematicians, and all physicists, would have been stupefied had it turned out otherwise. What was surprising about the proof is that Kepler's conjecture, which had resisted the efforts of mathematicians for four centuries, could be solved by nothing more advanced than linear methods. The series of papers, *Sphere Packings I, II, III, IV, V,* and *VI,* filled over 250 pages. The programs and the output, which can be found on Hales's web site, comprise 3 gigabytes of data.

APPENDIX TO CHAPTER 15

Some Mathematical Conjectures

The following list of 116 conjectures represents just a sample of challenges that mathematicians are confronted with. No attempt has been made at completeness. Some of the conjectures have already been solved, some have been shown to be incorrect, others may be solved by the time you read this. And, of course, new conjectures are proposed every year.

Adam's conjecture, Alperin's conjecture, Andrew-Curtis conjecture, annulus conjecture, Artin's conjecture, Banach conjecture, Bernstein's conjecture, Birch and Swinnerton-Dyer conjecture, Birch-Tate conjecture, Bombieri-Dworke conjecture, Borsuk's conjecture, Brauer's k(B) conjecture, Brauer-Thrall conjecture, Bunyakowski's conjecture, Burnside conjecture, C^1 stability conjecture, Calabi's conjecture, Carmichael's conjecture, Catalan-Dirkson conjecture, Catalan's conjecture, Chang's two cardinals conjecture, Cherlin-Zilber conjecture, Collatz conjecture, Conner-Floyd conjecture, Dyson's conjecture, Eckmann-Ruelle conjecture, entropy conjecture, epsilon conjecture, Erdös–Heilbronn conjecture, Erdös–Wintner conjecture, Evans' conjecture, Fejer's conjecture, Fermat's conjecture, four colour conjecture, Frobenius conjecture, fundamental conjecture of combinatorial topology, Gilbert-Pollak's conjecture, Goldbach's conjecture, Golod-Gulliksen conjecture, conjecture A of Golomb, Gottschalk's conjecture, Grothendieck's conjecture, Hadamard's conjecture, Hadwiger's conjecture, Halberstam's conjecture, Heawood's conjecture, Hedetniemi's conjecture, Hodge's conjecture, Iwasawa–Gleason conjecture, Kazhdan-Lustig conjecture, Kellogg's conjecture, Kelly-Ulam reconstruction conjecture, Kneser–Tits conjecture, generalized Kostrikin-Shafarevich conjecture, original Kostrikin-Shafarevich conjecture, Kummer's conjecture, Landau's conjecture, Leopoldt's conjecture, Lichnerowicz's conjecture, Lindelöf's conjecture, Luzin's conjecture, Macdonald's conjecture, MacWilliams-Sloane conjecture, Mahler's conjecture, Minkowski's conjecture, modularity conjecture, Moore space conjecture, Mordell's conjecture, Morley's conjecture, Mumford's conjecture, Nagata's conjecture, Nevanlinna's conjecture, Noether's conjecture, Novikov's conjecture, P. A. Smith's conjecture, $P \neq NP$ conjecture, Palis–Smale conjecture, perfect sequence conjecture, Petersson's conjecture, Pillai's conjecture, Platonov's conjecture, Poincaré's conjecture, positive mass conjecture, Ramanujan's conjecture, Reifenberg's conjecture, Riemann's conjecture, Schanuel's conjecture,

Schläfli's conjecture, Schönflies conjecture, Segal's conjecture, Selberg's conjecture, Serre's conjecture, Shapiro's conjecture, Shfarevich's conjecture, Siegel's conjecture, stable homeomorphism conjecture, Stone-Weierstrass conjecture, Strong Norikov conjecture, Suslin's conjecture, Szpiro's conjecture,[1] Tate's conjecture, the unicity conjecture, upper bound conjecture, van der Waerden's conjecture, Wagner's conjecture, Weil's conjecture, Weil-Tamiyama conjecture, Weinstein's conjecture, Witten's conjecture, the {XYZ} conjecture, Zeeman's conjecture.

---

[1] No relation to the author.

Bibliography

BOOKS ON SPHERE PACKINGS

Aste, T., and D. Weaire. *The Pursuit of Perfect Packing.* Philadelphia: Institute of Physics, 2000.

Conway, John H., and N. J. A. Sloane. *Sphere Packings, Lattices and Groups,* 3rd edition. Heidelberg: Springer-Verlag, 1999.

Fejes-Tóth, Laszlo. *Lagerungen in der Ebene, auf der Kugel und im Raum.* Heidelberg: Springer-Verlag, 1953.

Hsiang, Wu-Yi. *Least Action Principle of Crystal Formation of Dense Packing Type and Kepler's Conjecture.* Singapore: World Scientific Publishing Company, 2001.

Leppmeier, Max. *Kugelpackungen von Kepler bis heute.* Braunschweig: Vieweg, 1997.

Melissen, J. B. M., *Packing and Covering with Circles.* Ph.D. thesis. Utrecht, 1997.

Rogers, Carl A. *Packing and Covering.* Cambridge: Cambridge University Press, 1964.

Zong, Chuanming. *Strange Phenomena in Convex and Discrete Geometry.* Heidelberg: Springer-Verlag, 1996.

————. *Sphere Packings.* Heidelberg: Springer-Verlag, 1999.

OTHER BOOKS

Bak, Per. *How Nature Works,* New York: Copernicus, 1996.

Baumgardt, Carola. *Johannes Kepler: Life and Letters,* London: Gollancz, 1952.

Bühler, W. K. *Gauss: A Biographical Study,* Heidelberg: Springer-Verlag, 1981.

Conway, J. H., and F. Y. Fung. *The Sensual (Quadratic) Form,* Washington, D.C.: Mathematical Association of America, 1997.

Coxeter, H. M. S. *Introduction to Geometry,* New York: John Wiley & Sons, Inc., 1961.

Encyclopedia of Mathematics. Dordrecht: Kluwer Academic Publishers, 1997.

Fejes-Tóth, Laszlo. *Reguläre Figuren,* Budapest: Verlag der ungarischen Akademie der Wissenschaften, 1965.

Fuller, R. Buckminster. *Synergetics,* New York: Macmillan, 1975.

Gruber, P. M., and J. M. Wills. *Handbook of Convex Geometry,* Amsterdam: North Holland, 1993.

Grunbaum, Branko, and G. C. Shepherd. *Tilings and Patterns: An Introduction,* New York: W. H. Freeman and Company, 1986.

Hammer, Franz (Ed.). *Johannes Kepler: Selbstzeugnisse,* Stuttgart-Bad Cannstadt: Friedrich Frommann Verlag, 1971.

Kepler, Johannes. *The Six-Cornered Snowflake,* Oxford: Clarendon Press, 1966.

Keyes, J. Gregory, *Newton's Cannon,* New York: Random House, 1998.

Keynes, John M. *Essays in Biography,* London: Rupert Hart-Davis, 1951.

Koestler, A. *The Sleepwalkers* (first published 1959), New York: Viking Penguin, 1990.

Koza, John R. *Genetic Programming,* Cambridge, Mass.: MIT Press, 1992.

Meschkowski, Herbert. *Ungelöste und unlösbare Probleme der Geometrie,* Braunschweig: Vieweg, 1969.

Minkowski, Hermann. *Gesammlete Abhandlungen,* Volume 1, Leipzig: Teubner, 1911.

More. Louis T. *Isaac Newton: A Biography,* New York: Scribner, 1934.

Nagell, Trygve, Atle Selberg, Sigmund Selberg, and Knut Thalberg. *Selected Mathematical Papers of Axel Thue,* Oslo: Universitetsforlaget, 1977.

Rassias, George M. (Ed.). *The Mathematical Heritage of C. F. Gauss,* Singapore: World Scientific Publishing, 1991.

Rukeyser, Muriel. *The Traces of Thomas Hariot,* New York: Random House, 1971.

Scharlan W., and H. Opolka. *From Fermat to Minkowski,* Heidelberg: Springer-Verlag, 1985.

Serret, J.-A. *Oeuvres de Lagrange,* Volume 3, Paris: Gauthier-Villars, 1869.

Shirley, John W. (Ed.). *Thomas Harriot, Renaissance Scientist,* Oxford: Clarendon Press, 1974.

Siegel, Carl Ludwig. *Lectures on Quadratic Forms,* Bombay: Tata Institute of Fundamental Research, 1957.

―――. *Lectures on the Geometry of Numbers,* Heidelberg: Springer-Verlag, 1989.

Turnbull, H. W., J. F. Scott, and A. R. Hall. *The Correspondence of Isaac Newton,* 7 volumes, Cambridge: Cambridge University Press, 1959–1977.

Tymoczko, Thomas (Ed.). *New Directions in the Philosophy of Mathematics: An Anthology,* Princeton: Princeton University Press, 1998.

JOURNAL ARTICLES AND CHAPTERS USED OR MENTIONED

Barlow, William. "Probable Nature of the Internal Symmetry of Crystals." *Nature* (December 20, 1883): 186–188.

Barlow, William, and William Jackson Pope. "The Relation Between the Crystalline Form and the Chemical Constitution of Simple Inorganic Substances." *Journal of the Chemical Society* 91 (1907): 1150–1214.

Bender, C., "Bestimmung der grössten Anzahl gleich grosser Kugeln, welche sich auf eine Kugel von demselben Radius, wie die übrigen, auflegen lassen." *Archiv der mathematik und Physik* 56 (1874): 302–306.

Bezdek, Károly. "Isoperimetric Inequalities and the Dodecahedral Conjecture." *International Journal of Mathematics* 8 (1997): 759–780.

Blichfeldt, Hans F. "The Minimum Value of Quadratic Forms and the Closest Packing of Spheres." *Mathematische Annalen* 101 (1929): 605–608.

Boerdijk, A. H. "Some Remarks Concerning Close-Packing of Equal Spheres." *Philips Research Reports* 7 (1952): 303–313.

Cipra, Barry. "Gaps in a Sphere-Packing Proof?" *Science* 259 (12 February 1993): 895.

―――. "Music of the Spheres." *Science* 251 (1991): 1028.

―――. "Packing Challenge Mastered at Last." *Science* 281 (28 August 1998): 1267.

————. "Rounding out Solutions to Three Conjectures." *Science* 287 (17 March 2000): 1910–1911.

Coxeter, H. S. M. "An Upper Bound for the Number of Equal Nonoverlapping Spheres That Can Touch Another of the Same Size" in *Proceedings of Symposia in Pure Mathematics* 7, Providence: American Mathematical Society (1963): 53–71.

Dewar, Robert. "Computer Art: Sculptures of Polyhedral Networks Based on an Analogy to Crystal Structures Involving Hypothetical Carbon Atoms." *Leonardo* 15 (1982): 96–103.

Dold-Samplonius, Yvonne. "Interview with Bartel Leendert van der Waerden." *Notices of the American Mathematical Society* 44 (March 1997): 313–320.

Elkies, Noam D. "Lattices, Linear Codes, and Invariants." *Notices of the American Mathematical Society* 47 (November and December 2000): 1238–1245 and 1382–1391.

Fejes-Tóth, Laszlo, "Über einen geometrischen Satz." *Mathematische Zeitschrift* (1940).

————. "Über die dichteste Kugellagerung." *Mathematische Zeitschrift* (1943).

————. "Über dichteste Kreislagerungen und dünnste Kreisüberdeckungen." *Commentarii Mathematici Helvetici* 23 (1949): 342–349.

————. "Remarks on the Closest Packing of Convex Discs." *Commentarii Mathematici Helvetici* 53 (1978): 536–541.

Fejes-Tóth, Gabor, and W. Kuperberg. "Blichfeldt's Density Bound Revisited." *Mathematische Annalen* 295 (1993): 721–727.

————. "Packing and covering with convex sets." Chapter 3.3, Vol. B, in P. Gruber and J. Wills (Eds.), *Handbook of Convex Geometry,* Amsterdam: North-Holland, 1993.

Ferguson, Samuel P. "Sphere Packings V." Preprint (1997).

Frank, F. C. "Descartes' Observations on the Amsterdam Snowfalls of 4, 5, 6 and 9 February 1634." *Journal of Glaciology* 13 (1974): 535.

Freedman, David H. "Round Things in Square Spaces." *Discover* (January 1992): 36.

Gabai, David, G. Robert Meyerhoff, and Nathaniel Thurston "Homotopy Hyperbolic 3-Manifolds Are Hyperbolic." *Annals of Mathematics* (in press).

Gauss, Carl Friedrich. "Recension der 'Untersuchungen über die Eigenschaften der positiven ternären quadratischen Formen' von Ludwig August Seeber." *Göttingische gelehrte Anzeigen* 108 (July 9, 1831), and *Journal für die reine und angewandte Mathematik* (1840): 312–320.

Goldberg, David. "What Every Computer Scientist Should Know About Floating Point Arithmetic." *Computing Surveys* (March 1991).

Gregory, David, "Notebooks." Christ Church: manuscript number 131.

Guenther, S. "Ein stereometrisches Problem." *Archiv der Mathematik und Physik* 57 (1875): 209–215.

Hales, Thomas C. "The Sphere Packing Problem." *Journal of Computational and Applied Mathematics* 44 (1992): 41–76.

————. "Remarks on the Density of Sphere Packings in Three Dimensions." *Combinatorica* 13 (1993): 181–197.

————. "The Status of the Kepler Conjecture." *The Mathematical Intelligencer* 16 (1994): 47–58.

————. "Sphere Packings I." *Discrete and Computational Geometry* 18 (1997): 135–149.

————. "Sphere Packings II." *Discrete and Computational Geometry* 17 (1997): 1–51.

————. "A Formulation of the Kepler Conjecture." Preprint (1998).

————. "The Kepler Conjecture." Preprint (1998).

————. "An Overview of the Kepler Conjecture." Preprint (1998).

————. "Sphere Packings III." Preprint (1998).

————. "Sphere Packings IV." Preprint (1998).

————. "Cannonballs and Honeycombs." *Notices of the American Mathematical Society* 47 (April 2000): 440–449.

Hales, Thomas C., and Sean McLaughlin. "A Proof of the Dodecahedral Conjecture." Preprint (1998).

Hargittai, István. "Lifelong Symmetry: A Conversation with H. M. S. Coxeter." *The Mathematical Intelligencer* 18 (1996): 35–41.

Hérmite, Charles. "Sur la réduction des formes quadratiques ternaires" in *Oeuvres III,* Paris: Gauthiers-Villars (1908).

————. "Sur la théorie des formes quadratiques ternaires" in *Oeuvres 1* Paris: Gauthiers-Villars (1905).

Hilbert, David. "Mathematische Probleme." *Archiv der Mathematik und Physik* 1 (1901): 44–63 and 213–237.

Hoppe, Reinhold. "Bemerkung der Redaction." *Archiv der Mathematik und Physik* 56 (1874): 307–312.

Horgan, John. "The Death of Proof." *New Scientist* (May 8, 1993): 74–82.

Hsiang, Wu-Yi. "On the Sphere Packing Problem and the Proof of Kepler's Conjecture." *International Journal of Mathematics* 4 (1993): 739–831.

————. "A Rejoinder to Hales's Article." *The Mathematical Intelligencer* 17 (1995): 35–42.

Kantor, Jean-Michel. "Hilbert's Problems and Their Sequels." *The Mathematical Intelligencer* 18 (1996): 21–30.

Kershner, Richard. "The Numbers of Circles Covering a Set." *American Journal of Mathematics* 61 (1939): 665–671.

Klarreich, Erica. "Foams and Honeycombs." *Scientific American* 88 (March/April 2000): 152–161.

Kleiner, Israel, and Nitsa Movshovitz-Hadar, "Proof: A Many Splendored Thing." *The Mathematical Intelligencer* 19 (1997): 16–26.

Kolata, Gina. "Scientist at Work: John H. Conway." *The New York Times* (October 12, 1993).

Lagarias, J. C. "Local Density Bounds for Sphere Packings and Kepler's Conjecture." Preprint (1999).

Lam, C. W. H. "How Reliable Is a Computer-Based Proof?" *The Mathematical Intelligencer* 12 (1990): 8–12.

Lampe, E. "Nachruf für Reinhold Hoppe." *Archiv der Mathematik und Physik* 1 (1900), 4–19.

Lebesgue, V. A. "La réduction des formes quadratiques définie positives à coefficients réels quelconques, démonstration du théorème de Seeber sur les réduites des formes ternaires." *Journal de Mathématiques Pures et Appliquées* Série 2, Volume 1 (1956).

Leech, John. "The Problem of the Thirteen Spheres." *The Mathematical Gazette* 40 (1956): 22–23.

Lindsey, J. H. Jr. "Sphere Packing in R^2." *Mathematika* 33 (1986): 137–147.

Logothetti, Dave, "H. S. M. Coxeter." in Albers and Anderson (Eds.), *Mathematical People,* Basel: Birkhäuser (1985).

Lüthy, Christoph, "Bruno's *Area Democriti* and the origins of atomist imagery." *Bruniana and Campanelliana* 1 (1998): 59–92.

———. "The invention of atomist iconography." Preprint 141, Max-Planck Institut für Wissenschaftsgeschichte (2000).

MacLane, Saunders. "Mathematics at Göttingen Under the Nazis." *Notices of the American Mathematical Society* 42 (October 1995): 1134–1138.

———. "Van der Waerden's Modern Algebra." *Notices of the American Mathematical Society* 44 (March 1997): 321–322.

Mahler, K. "On Reduced Positive Definite Ternay Quadratic Forms." *Journal of the London Mathematical Society* 15 (1940): 193–195.

Melmore, Sidney. "Densest Packing of Equal Spheres." *Nature* (June 14, 1947): 817.

Milnor, John. "Hilbert's Problem 18: On Crystallographic Groups, Fundamental Domains, and on Sphere Packing." *Proceedings of Symposia in Pure Mathematics* 28 (1976): 491–506.

Möhring, Willi. "Hilbert's 18th Problem and the Göttingen Town Library." *The Mathematical Intelligencer* 20 (1998): 43–44.

Muder, Douglas J. "A New Bound on the Local Density of Sphere Packings." *Discrete and Computational Geometry* 10 (1993): 351–375.

———. "Putting the Best Face on a Voronoi Polyhedron." *Proceedings of the London Mathematical Society* 56 (1988): 329–358.

Oler, N. "An Inequality in the Geometry of Numbers." *Acta Mathematica* 105 (1961): 19–48.

Oppenheim, A. "Remark on the minimum quadratic form." *Journal of the London Mathematical Society* 21 (1946): 251–252.

Peli, Gabor, and Bart Noteboom. "Market Partitioning and the Geometry of the Resource Space." *American Journal of Sociology* 104 (1999): 1132–1153.

Phillips, Ralph. "Reminscences About the 1930s." *The Mathematical Intelligencer* 16 (1994): 6–8.

Pohlers, Wolgang. "In Memoriam: Kurt Schütte 1909–1998." *The Bulletin of Symbolic Logic* 6 (2000): 101–102.

Rankin, R. A. "On the Closest Packing of Spheres in n Dimensions." *Annals of Mathematics* 48 (1947): 1062–1081.

Rogers, Carl A. "The Packing of Equal Spheres." *Proceedings of the London Mathematical Society* 8 (1958): 609–620.

Rousseau, G. "On Gauss's Proof of Seeber's Theorem." *Aequationes Mathematicae* 43 (1992): 145–155.

Sangalli, Arturo. "The Easy Way to Check Hard Maths." *New Scientist* (October 1993).

Schütte K., and B. L. van der Waerden. "Das Problem der dreizehn Kugeln." *Mathematische Annalen* 125 (1953): 325–334.

Segre, B., and K. Mahler. "On the Densest Packing of Circles." *American Mathematical Monthly* 51 (1944): 261–270.

Seiden, Steve. "Can a Computer Proof Be Elegant?." Preprint (2000).

———. "A Manifesto for the Computational Method." Preprint (October 2000).

Seife, Charles. "Mathemagician" *The Sciences* (May/June 1994): 12–15.

Severance, Charles. "An Interview with the Old Man of Floating Point." *IEEE Computer* (March 1998).

Singh, Simon. "Packing Them In." *New Scientist* (June 28, 1997).

Sloane N. J. A. "The Packing of Spheres." *Scientific American* 250 (January 1984): 116–125.

———. "The Sphere Packing Problem." *Documenta Mathematica* (1998).

Solomon, Ron. "On Finite Simple Groups and Their Classification." *Notices of the American Mathematical Association* 32 (February 1995): 231–239.

Stewart, Ian, "Has the Sphere Packing Problem Been Solved?" *New Scientist* (May 1992): 16.

———. "The Kissing Number." *Scientific American* (February 1992): 90–92.

Swart, E. R. "The Philosophical Implications of the Four-Color Problem." *American Mathematical Monthly* 87 (1980): 697–707.

Szpiro, George G. "Cycles and Circles in Roundoff Errors." *Physical Review E* 47 (1993): 4560–4563.

———. "Forecasting Chaotic Time-Series with Genetic Algorithms." *Physical Review E.* 55 (1997): 2557–2568.

Thue, Axel. "Dichteste Zusammenstellung von kongruenten Kreisen in einer Ebene." *Kra. Vidensk. Selsk. Skrifter I. Mat. Nat. Kl.* (1910).

———. "Om nogle geometrisk-taltheoretiske Theoremer." *Forh. Ved de skandinaviske naturforskeres* (1892): 352–353.

Thurston, William P., "On Proof and Progress in Mathematics." *Bulletin of the American Mathematical Society* 30 (April 1994): 161–177.

Torquato, S., T. M. Truskett, and P. G. Debenedetti. "Is Random Close Packing of Spheres Well Defined." *Physical Review Letters* 84 (March 2000): 2064–2067.

Wills, J. M. "Finite Sphere Packings and the Methods of Blichfeldt and Rankin." *Acta Mathematica Hungarica* 75 (1997): 337–342.

Zeilberger, Doron, "Theorems for a Price: Tomorrow's Semi-Rigorous Mathematical Culture." *Notices of the American Mathematical Society* 40 (October 1993): 978–981.

Index